FRERE APOGEE SURE

SOLDIER'S WATCH KONTADOR

ERMETO-BAG PULLMAN

OPLAN ACVATIC CRONOPLAN

CALENDOGRAF CALENDOMATIC

CELESTOGRAF CALENDOSCOPE

IC ASTRONIC TEMPOMATIC

UTOMATIC 331 FUTURAMIC

JM WATCH SUB-SEA DATRON

MOVADO

Fritz von Osterhausen

THE MOVADO HISTORY

Schiffer Publishing Ltd

77 Lower Valley Road, Atglen, PA 19310

Acknowledgements

This book could not have been realised without the initial decision taken by NAWC's executive officers Gedalio and Efraim Grinberg to produce it, or without their generous support as well as that of their associates Bernhard Stoeber and Chuck Davidson in Lyndhurst and Kurt Burki, Florian Strasser and Denise Hofmann in Bienne. I would like to make a special mention of Bernhard Stoeber, who constantly accompanied my work in a critical, constructive and encouraging manner, collecting such a mass of documents, photographs and watches that one could almost say I had provided him with support rather than the other way round. Also, I would like to extend my special thanks to Chuck Davidson for his significant contributions to the layout and design of this book.

I am moreover extremely grateful to the following ladies, gentlemen and institutions for their assistance.

George Mauro, Little Falls, N.J., for his excellent photos.

Matthew Bain from Miami, Ike Bernstein from New Jersey, Jean-Marc Barrelet from Neuchâtel town archives, Bernard Ditesheim from La Chaux-de-Fonds and Gérard Ditesheim from Greenwich, CT., Deanne Torbert Dunning and Edward Faber from New York, Henry B. Fried from Larchmont, N.Y., Herbert Neumüllers from Heidelberg, J.-M. Piguet from MIH in La Chaux-de-Fonds, Kathleen H. Pritchard from Bethseda, MD., Melissa Seiler from Brooklyn Museum, New York, Steven van Dyk from the Cooper Hewitt Museum, New York, Antoine Simonin from Neuchâtel, Vivian Swift, formerly of Christie's, New York, Christian and Michael Pfeiffer-Belli, Gisbert L. Brunner and Gerhard Streitberger, all from Munich, as well as the following auction houses: Dorotheum in Vienna, Antiquorum in Geneva and Klöter in Schloss Dätzingen.

Also, I would like to thank Derek Pratt for his fine work in preparing the English translation and Debra Gibson for proofreading.

And, finally, this book may not be concluded without a vote of thanks to the many nameless employees of the Movado Company, who with their efforts have contributed and will continue contributing to keeping Movado "always in motion".

Fritz von Osterhausen

Copyright © 1996 by Movado Group, Inc.
Library of Congress Catalog Card Number: 96-068730.

The original German edition was published by Verlag Georg D. W. Callwey, Munich.

Translator: Derek Pratt; Balm, Switzerland

Layout and Production: Christian and Michael Pfeiffer-Belli, Munich.
Dust Cover Design: Charles Davidson, New York and HBC-Design, Munich.
Typesetting: Gerber Satz, Munich.
Lithos: Karl Findl, Icking near Munich.
Printing and Binding: Amilcare Pizzi, Cinisello-Balsamo, Milano.

Printed in Italy
ISBN 0-7643-0126-8

Published by Schiffer Publishing Ltd.
77 Lower Valley Road
Atglen, PA 19310
(610) 593-1777
Please write for a free catalog.
This book may be purchased from the publisher.
Please include $2.95 for shipping.
Try your bookstore first.

We are interested in hearing from authors with book ideas on related subjects.

Contents

Acknowledgements 4
Foreword 6
Introduction, the history of watchmaking
in Geneva and in the Swiss Jura 7

The Past

Early family history,
   La Chaux-de-Fonds in 1870 9
The beginning: The company's foundation
   in 1881 and the first years 14
1905: The birth of the Movado brand
   and its first years up to 1910 22
Patents: Early patents 1902–1912
   A summary of the patents up to 1969 26
1912: The "Polyplan" 34
Catalogues: 15 years of Movado
   between 1910 and 1925 39
   The movement calibres 47
   The First World War Military
   or Soldier's Watch 50
Art Deco: The 1920's:
   "Valentino" and "Ermeto" 66
1926: Ermeto 74
Chronometer trials I
   1899–1939: Pocket and deck watches . . . 90
   1930: Developments in the thirties 104
   1938: Wrist chronographs up to the
     forties 118
   1945: Wrist calendar watches up
     to the fifties 124
Automatic winding: The forties and
   fifties at Movado 128
Chronometer trials II
   1950–1969: Wrist chronometers 144
The Museum Watch:
   Nathan George Horwitt 148
1960: The path to "El Primero" 154
1969–1983: The years with Zenith 158

The Present

Acquisition and new start 164
Organisation 167
Artists' Watches: The new Movado models . 168
   Andy Warhol Times/5 168
   Yaacov Agam's "Rainbow" Collection 170
   Arman's "The color of time" 172
   James Rosenquist's
     "Elapse, Eclipse, Ellipse" 174
   Max Bill's "Bill-time" 176
   Romero Britto's
     "The Children of the World" 178
"Collection 1881" Replicas: 180
   Reproductions 182
   The First World War Soldier's Watch 182
   The Bauhaus Watch 182
   Further models 184
Specials: Famous owners of
   Movado watches 188

Appendices

Appendix 1: Movado numbering and
           reference systems 194
Appendix 2: Complete list of all
           Movado mechanical watch
           calibres 1910–1980 204
Appendix 3: Results of Movado pocket/deck
           watches at Neuchâtel and Kew/
           Teddington Observatories 216
Appendix 4: Observatory wrist
           chronometers 220
Appendix 5: Movado inventions and
           patents 222
Appendix 6: Swiss case makers 228
Appendix 7: Museums in which Movado
           watches are to be found 230
           Sources of illustrations 231
           Bibliography 231
           Index 232

Foreword

When first Mr Burki, and then Mr Stoeber asked me if I was willing to co-operate in a book project on the history of Movado, I accepted with enthusiasm.

Movado's history is closely linked to my own youth, my apprenticeship and finally my professional career.

One of my childhood memories is accompanying my father – who was a nephew of the founder of Movado and its technical director – to the factory on Sunday mornings, where he went to find out the latest; in the Ditesheim family, daily contact – even on Sundays – with "the factory" was a matter of course.

My brother and I were allowed to enter the workshop where the high precision adjuster worked. The silence in this room was interrupted only by the steady ticking of the seconds regulators used to rate watches precisely.

Later at the watchmaking school I became aware of the good reputation that Movado had at a time when handwork was still of prime importance. With about 350 employees, Movado was the most important company in La-Chaux-de-Fonds: with its own development office, manufacturing department for ébauches and components, and of course workshops for assembly, timing and final control. Very early, in the decade from 1910/20, when the market was still completely geared to pocket watches, Movado was advancing the development of wrist watch movements. Calibre 400, with which the wrist watch model "Polyplan" had been equipped since 1912, deserves a special mention here. In fact, the development of precise, reliable and flat movements, allowing the design of elegant watches, was one of Movado's main objectives.

At this point we should also mention "Ermeto", a pocket watch dating from 1926 with automatic winding, which was often copied but never equalled. Many of these pieces, which were ahead of their time, were produced into the 1980's and are today highly desirable collector's items.

Movado's participation in chronometer trials organised by observatories, particularly that of Neuchâtel, and the very good results attained there, testified to the company's high quality precision adjusting. First places in the wrist watch category, in the years 1956, 1957 and 1958, led to an improvement of all wrist watch calibres by increasing the balance frequency from 18 000 to 21 600 vibrations per hour and later, with the "Clinergic 21" escapement, up to 36 000 vibrations per hour.

The collaboration resulting from the transfer of the research work of the high precision adjusters to industrial production was decisive in improving the precision of Movado's watches as well as their reliability and longevity. This is a positive aspect of the rôle played by observatory trials with the knowledge thereby gained for everyday watches.

When, after receiving my technical college diploma in Neuchâtel, I came to New York in 1951 to gain further professional experience, I saw for myself the influence and the strong position of Movado's American branch both in the USA and on the whole American continent. The strong position of the American branch after the opening in 1924 is due to the intensive efforts of Gaston Ditesheim and his successors. Movado's strong position there is still the envy of many Swiss watch companies today.

Another important advertisement for Movado was the "Museum" watch. Introduced at the end of the 1950's, it became a worldwide success, and is as popular now as it was then. Today, Movado's dynamic management has equipped this model with a solar cell, thus cleverly combining scientific progress with the aesthetic appearance of the watch.

The meaning of the Esperanto word Movado – always in motion – is thus once more impressively apt.

Bernard Ditesheim

(Bernard Ditesheim is a nephew of Isaac the "engraver", who left his successful engraving workshop in 1905 to enter the company belonging to his three brothers Achille, Léopold and Isidore.)

Introduction, the history of watchmaking in Geneva and in the Swiss Jura

It is impossible to separate the history of the Swiss watch company Movado, the subject of this book, from Swiss watchmaking history. Movado is an achievement of the Ditesheim family, who founded this company and led it to greatness, running it for 88 years under family management. A trade with its fascinating end product that has determined life for two centuries in this small country, above all in the Swiss Jura region, to an extent and exclusivity seldom seen elsewhere.

The development of watchmaking began in Switzerland in the 16th century, somewhat later than in France and Germany, but at about the same time as in England. It was not until the late 18th century, however, that watchmaking reached the high level of importance it enjoys today. Up to then, the measurement of time and the corresponding division of the day into regular intervals had been completely unknown and also of little importance to the normal citizen. People oriented their lives according to the height of the sun and the succession of day and night. Or, at best, after the ringing of bells in public clock towers, already widely known in the 14th century. Until this time, the possession of a personal watch or clock, "personal time" was a privilege of the rich and powerful: the prince, the nobility and the high clergy.

Three very different locales, one town and two regions in Western Switzerland, saw the birth of the watchmaking trade: Geneva; the area around La Sagne in the Neuchâtel mountains; and the Joux Valley, with its main town of Le Brassus. The order given here is not accidental; it marks a chronological succession, as we will see.

In Geneva, the first portable timepieces – mainly descendants of the small table clocks with horizontal dials – had been made since about 1540, at first by French craftsmen. These early watchmakers were Huguenots, who had been persecuted in France for religious reasons and had therefore left their homes. The first watchmaker entered in the town register in 1554 was a Frenchman: Thomas Bayard, who was also a goldsmith.

In 1566, Calvin forbade the goldsmiths in Protestant Geneva to make sacred implements, such as crosses, chalices, etc. This ban represented a boom for watchmaking, since goldsmiths, in order to continue exercising their profession, used their skills in the manufacture of timepieces and their cases. In the same year of 1566, a watchmakers' guild was founded in Geneva, the second oldest after Paris (1544). In 1601, it passed regulations with strict provisions on the protection against inferior workmanship and unfair competition.

In 1685, following France's lifting of the Edict of Nantes, which had granted the Huguenots religious freedom, a new wave of French exiles came to Geneva. Among their number were many qualified artistic craftsmen – including watchmakers – and by the end of the 17th century watchmaking had become the most important trade in the city. Since then Geneva has been known as the centre of especially prestigious and precious objects – luxury watches.

At the end of the 17th century, when Geneva had long been an important centre of watchmaking, the trade was still largely unknown in the adjoining Canton of Vaud and in the Neuchâtel Jura. Only a few watchmakers were to be found in large towns such as Lausanne, Rolle and Neuchâtel. It is therefore possible that the almost legendary story about the young smith and mechanic from La Sagne, Daniel JeanRichard, actually took place in this region. One day, he is said to have received a visit from a traveller with a broken pocket watch and, although JeanRichard had never seen a portable mechanical timekeeper in his life, he had soon found out how it worked and why it was not working, subsequently repairing it with home-made tools. Then, as the story goes, JeanRichard felt an urge to learn watchmaking systematically. However, as he was unable to find a master craftsman in the Geneva region willing to reveal to him the secrets and skills of watchmaking, he taught himself everything, and went on to teach others. This is the story of how watchmaking came to La Sagne and to the Neuchâtel Jura.

Even if this story sounds rather like a fairy tale, there is documentary evidence for the existence of Daniel JeanRichard (1672 – 1741), who was the great pioneer in the production of watches in the Neuchâtel region. Obviously a man of great mechanical talent and almost superhuman tenacity, JeanRichard is reputed to have had a decisive influence on the development of a method of production based upon the division of labour. This system proved to be ideal for this rugged area with its far flung villages, hamlets and isolated farms.

This system entailed only a few components being made at home in family workshops, work which was done as a secondary job in winter, with farmwork as a priority in the warmer season. Travelling "établisseurs" collected the parts, assembled them in larger, central workshops, finished them, and finally put the complete watches on the market. In this way, watchmaking also started up in the two main towns of the Jura region after Neuchâtel: in Le Locle, where JeanRichard himself settled in 1700, and La Chaux-de-Fonds.

In the course of time Le Locle developed into a chronometry or high precision watchmaking centre. Abraham-Louis Breguet worked here during his forced period of asylum in the years 1793 – 1795. Other such famous chronometer makers drawn to the area included Jacques-Frédéric Houriet, Henri Grandjean, the Danish Jürgensens (Urban, Louis Urban, Jules and Jacques-Alfred), Frédéric-Louis Favre-Bulle and Ulysse Nardin.

La Chaux-de-Fonds, on the other hand, was the centre of simple but good quality everyday watches, produced in large series. The factory of Georg Friedrich Roskopf, with the reasonably priced "Roskopf" watches, was characteristic in this respect. However, this company also produced valuable high precision timepieces.

In the third centre of Swiss watchmaking – the Joux Valley – work started a generation later, shortly before Daniel JeanRichard's death in 1740. In the same year the lively young farmer's son, Samuel-Olivier Meylan, who was no longer satisfied with the meagre agricultural existence and winter unemployment in this isolated mountain valley, set out for the small town of Rolle on the Lake of Geneva. There he took up an apprenticeship with the master craftsman Mathieu Biaudet. On completing his training Meylan returned to his hamlet, Chez-le-Maître, and began making complete watches by hand. He thus started a horological avalanche in the Joux Valley, being followed by others who were just as dissatisfied with the limited opportunities of their rural isolation as Meylan had been. Following his example, they received a watchmaker's training – above all in Fleurier, where the guild regulations were less strict than in Rolle. Their example led to the gradual transformation of the Joux Valley from a largely poor agricultural region into a prosperous watchmaking centre, with farming as a subsidiary occupation. Later, after the Napoleonic era, when ébauches and complicated under-dial work from the Joux Valley were in such great demand in Geneva, Paris, and other centres that suppliers could hardly keep up with orders, people spoke of this time as the "Golden Age". As in Geneva, Le Locle and La Chaux-de-Fonds, the Joux Valley also specialised, namely in complicated pocket watches and under-dial work (additional mechanisms to the normal movement). The famous masters responsible for this highly complicated work were only known to insiders at that time, as they seldom signed their tiny miracles themselves; this was done by the companies which marketed the finished products. There were watchmakers such as Georges Golay, Louis Audemars, Louis Elisée Piguet, Victorin Piguet, Daniel Aubert and Charles-Antoine LeCoultre, each from long-established families in the region. In fact, it is characteristic of the early isolation of the Joux Valley that watchmaking and even public life itself were dominated well into the 20th century by the members of ten families who had been settled there since the Middle Ages. These families often intermarried and were constantly forming new business connections with one another.

The attentive reader, who is surely surprised that it took a period of far more than 150 years for a trade such as watchmaking to spread in a limited region like the Swiss Jura and to establish itself in the Joux Valley only 50 km from Geneva, should realise how unimaginably isolated and inaccessible this valley was in former times. Nowadays it hardly takes a comfortable half hour by car over good roads from Geneva to Le Brassus, but this valley remained secluded and inaccessible far into the 19th century, although intensive business relations with Geneva and other centres had existed for a long time. Business connections which were also to become important for Movado a few decades later, as this firm had, for its part, already at an early stage formed good relations with the centres beyond the nearby Neuchâtel mountains.

Early family history, La Chaux-de-Fonds in 1870

1a
The Grande Rue (later Rue Léopold-Robert) in La-Chaux-de-Fonds in 1867.

The year 1876 was a relatively calm one at international level. It was in this year that the English Queen Victoria adopted the title "Empress of India", the American writer Jack London was born and the French poetess George Sand died. In the same year, Alexander Bell developed the first usable form of the telephone and there was a great famine in Northern China, lasting until 1878. In the year 1876, a delegation of officials from the Swiss watch industry travelled to the World Exhibition in Philadelphia, USA. There, they were horrified to learn that, in the space of a few years, the Americans had built up a very efficient, modern, factory-type mass production of pocket watches. With the enormous number of high quality and reasonably priced pocket watches manufactured, the Americans were perceived as serious competition for Switzerland. The report of the returning delegates caused disbelief and even consternation in some circles, setting off a nationwide discussion on how this challenge was to be met.

In 1876, a tinter of engravings and probably an inn-keeper named Samuel Ditesheim from Hegenheim in the then Prussian province of Alsace, left home with his wife Thérèse and six of their seven sons to emigrate to La Chaux-de-Fonds in the Swiss Jura. Hegenheim lies directly on the French-Swiss border near Basel, and could be considered a suburb of Basel, were there not a border between them.

The Ditesheims' reasons for emigrating are unknown. Perhaps it was the fascination of La Chaux-de-Fonds as a watchmaking centre. Maybe the uncertain political situation following the Franco-Prussian War of 1870/71 was also partly responsible for the serious step taken by the Ditesheims and many others at that time. The annexation of Alsace-Lothringen after defeat in this war gave the French no peace. France had already begun to rearm and the possibility of a new war was openly discussed. In 1875, Bismarck had warned France about their rearmament and the term "war in sight" was often in people's mouths. A war whose focal point and battlefield would have been Alsace, which France wished to recapture.

In any event there must have been serious reasons for the Ditesheims' decision to leave Hegenheim, as the father Samuel, born in 1812 and now aged 64, was at an age when people are more likely to retire than start a new life – unless it is unavoidable.

Their emigration seems to have been well planned, as the eldest son Abraham (1858–1877), who had been trained as an engraver, came to the Swiss Jura several years before as the advance guard, as the "billeting officer", to settle in La Chaux-de-Fonds and open an engraver's workshop. Engraving, mainly of pocket watch cases, was a popular and desirable profession in the watchmaking centre of La Chaux-de-Fonds at that time; in 1895 there were over 600 engraving workshops, a number which clearly shows that the town was a worldwide centre of the engraver's handwork in that era.

Abraham died young, aged only 19. He had set off bravely for the unknown at an early age – 16 years old at the most – after a short apprenticeship, perhaps at his father's wish. In 1876, a year before his premature death, he had been joined by the rest of the family. This group consisted of father Samuel, his wife Thérèse, 21 years younger than her husband, and their six sons. Louis, the youngest, died at the age of 13. The others all learned the professions connected with watchmaking: Léopold (1860 – 1933), Achille (1862 – 1944) and Isidore (1868 – 1941) became watchmakers, Isaac (1860 – 1928) was an engraver like Abraham, and Aron (1864 – 1944) became an engine-turner.

For this family of Jewish emigrants, the new circumstances in La Chaux-de-Fonds cannot have been easy, as the negative attitude of the locals towards exiled Jews is documented. Presumably the situation was mitigated by the fact that many of their fellow-Jews were active in the watchmaking and watch-selling trades. There were, for example, Achille Hirsch, Roskopf's partner Charles-Léon Schmid, Moise Dreyfuss with the Rotary and Enila Companies, the Pery Watch Co. belonging to the Dreyfus family, Alphonse Braunschweig, who had founded the watchmaking factory Election in 1848. And finally their numerous namesakes, all from Hegenheim, above all the brothers Marc and Emanuel Didisheim, founders of the Marvin Company, who had emigrated to nearby St. Imier in 1850 and had moved their factory to La Chaux-de-Fonds in 1894.

And then there were the Ditisheims, who had already come to La Chaux-de-Fonds from Hegenheim in 1834. Like the Ditesheims, the family of Gaspard Ditisheim also reflected the situation in this region at that time, dominated as it was by watchmaking. Four of his five sons became watchmakers, the fifth an engraver. We only mention this because one of the five sons was to become one of the most famous watchmakers: this was Paul Ditisheim, born in 1868. A quotation from Mark Perrenoud's work (Un rabbin dans la cite, Jules Wolff, Neuchâtel 1989) clearly describes the social situation mentioned above:
"At the beginning of the (20th) century, La Chaux-de-Fonds' advantage over other centres in the Jura was largely due to the presence of an Israelite community, which knew how to provide this town with impulses exceeding those of a purely economic nature."

There was then an active Jewish community that had played an important rôle in the growth of the watchmaking industry, and which largely consisted of Frenchmen from the Alsace, who could offer immigrants like the Ditesheims some support in their difficult beginnings. This solidarity among the members of the Jewish community of La Chaux-de-Fonds is also clearly demonstrated by the fact that three of the brothers (Léopold, Isaac and Achille) married three sisters from the Jewish Lévy family, who had emigrated to Switzerland from the Vosges town of St. Dié. Incidentally, Thérèse Ditesheim, their mother, had also been a Lévy before her marriage.

As we have already digressed somewhat, let us take a closer look at the town La Chaux-de-Fonds, to which Samuel Ditesheim and his family had decided to emigrate.

La Chaux-de-Fonds had originated in a mountain valley situated 1,000 m above sea level in the Swiss Jura, not far from the border with France. It first appears in documents dating from the mid-14th century, when some colonists from the Val-de-Ruz settled there. Its definitive present-day form was established during the first half of the nineteenth century after the town fire of 1794. We find a strict grid system of streets with regular, longitudinal main axes and sharply rising subsidiary axes running crosswise to these, around which the high street, Avenue Léopold Robert, forms the central longitudinal axis at the lowest point, somewhat like a backbone. The generosity and spaciousness of the town, combined with the stateliness of the houses, which were planned according to a uniform concept, show that the town was already wealthy at the time of its reconstruction – or that its administration firmly expected it to become so. This was not quite a matter of course, for the poor economic climate in this mountain valley did not favour a flourishing and lucrative agriculture. As in other parts of the Jura region, the Middle Ages already saw the arrival of the charcoal burner and, as a consequence, iron production and smelting, there being iron ore deposits. In the course of the centuries, the professions of specialised metal

La Chaux-de-Fonds. But also the production of clocks continued to be located here, especially the manufacture of pendulum clock cases. The cabinet maker Abraham-Louis Sandoz from La Chaux-de-Fonds was one of the first Swiss crafts-men to specialise in the decoration of pendulum clock cases, as he mainly worked for Neuchâtel clockmakers.

The development, or rather the expansion

1 b
View of La Chaux-de-Fonds, about 1900.

workers developed, such as the blacksmith, nailmaker and gunsmith, whose more intricate variants, such as the goldsmith's craft, all led to clockmaking.

Watchmaking developed gradually at the end of the 17th century on the basis of Daniel Jean Richard's activities in the Neuchâtel region. The division of labour system introduced by him there for portable timepieces led particularly in La Chaux-de-Fonds to a great number of small, spe-cialised workshops, which staunchly held their own longer than elsewhere, even well into the era of factory production. In fact, they still have a cer-tain status in the town today.

JeanRichard, as mentioned above, had organised the production of a miniature clock called the pocket watch. By the mid-18th century, this had become an important economic factor in

of La Chaux-de-Fonds in the 19th century, can best be described as hectic. Besides the expanding watch industry, there was also the establishment of a railway connection between Le Locle and Neuchâtel in 1860. The population increased from 2,266 in 1760 to over 16,000 in 1860. In 1879, 21,887 inhabitants were counted, and shortly be-fore the First World War, there were over 40,000. More than 500,000 watches were made in this live-ly industrial town in 1853; all of them still largely made in the division of labour system in numer-ous, small specialised workshops. Hundreds of chimneys must have smoked there each day, as the watchmaker needed a heated room for his delicate work in winter.

Extensive investments were also made in the field of education. In 1855, an industrial school was founded, offering secondary school perhaps there was also some family connection to the Didisheims and the Ditisheims, as the different spellings are too slight to be accidental. Moreover, the father Samuel was not only an innkeeper, as passed down in the family, but also a tinter. We know this from a copy of Isidore Ditesheim's birth certificate, made out by the registrar in Hegenheim in 1895. In this document the father is described as "Dietisheim Samuel Cerf (Hirsch), tinter". The spelling of the family name with "ie" is also interesting. A tinter was someone who painted engravings or maybe woodcuts, and the profession of the tinter was considered an artistic one. Thus it is not surprising that a tinter's sons should be interested in the related art of the engraver, and does it not seem likely that they decided to learn and practise this in the nearby world centre of the engraver's art, La Chaux-de-Fonds? However, the eldest son Abraham must, judging from his age, already have made his personal decision about a profession and his engraver's training, and acted upon this, before emigrating. Of Samuel's remaining five sons, Léopold and Isaac were 16 years old at this time and Achille 14. In the case of the first two, the decision to become a watchmaker and the other an engraver could have been taken before emigration. The same could also be true for Achille. (The famous Paul Ditisheim, for example, began his training at the watchmaking school in La Chaux-de-Fonds at the age of 13). The wide palette of opportunities for training at a school in La Chaux-de-Fonds must have exercised a great attraction on all Samuel Ditesheim's five sons, whether they had already begun their training beforehand or not. For in La Chaux-de-Fonds, the traditional apprenticeship practised elsewhere was gradually being superseded by the training at the industrial, the watchmaking and the commercial schools. At the watchmaking school, it was already possible to learn the mechanics required for watch manufacture in a special class, established in 1876. And it was just the fundamentals of mechanisation that would have formed such a valuable basis for Achille Ditesheim's future career.

education for 12–16 year olds. This was followed by a watchmaking school in 1865, the second one in Switzerland after the "Ecole des Blancs", established in Geneva in 1824. There was a commercial school, and later the "Chambre d'Horlogerie suisse", the Swiss Chamber of Horology, was established there.

By 1850, La Chaux-de-Fonds had become the centre of the Swiss watch industry, having seemingly overtaken Geneva, which had been given this title by the Swiss before. Many people even somewhat euphorically described the town as the "métropôle mondiale de l'horlogerie", the world's watchmaking capital. The Ditesheims had come to this lively, modern industrial town in the mountains and were granted Swiss citizenship surprisingly fast, even in the year of their arrival in 1876. Not, however, in La Chaux-de-Fonds itself; they were obliged to "purchase" the right of domicile in the nearby village, Cerneux-Péquignot. Already at that time, Swiss citizenship was a much desired and difficult goal, with most immigrants having to wait for many years to attain it.

As mentioned, it cannot be ascertained today why the Ditesheims decided to emigrate to La Chaux-de-Fonds and to completely adopt watchmaking and its related professions. However, the circumstances made this decision plausible, once emigration had been decided on.

First of all, there was the example of their namesakes from the same village Hegenheim;

The company's foundation in 1881 and the first years

4
The building at Rue 1er Mars, No. 13, in La Chaux-de-Fonds, where Achille Ditesheim's watch company was established.

5
An interior view of a watchmaking workshop shown at the National Exhibition in Zurich in 1883.

In 1881, just five years after his family had arrived in La Chaux-de-Fonds, Achille Ditesheim, who must have just finished his training at the watchmaking school, founded his own watchmaking company. He first established this together with 6 workmen in the house at Rue 1er Mars 13, which is still standing today. It is one of those typical three-storey plastered houses with stone-work at its corners, a high-level basement and hipped roof, as were built in their hundreds following the town fire of 1794, in strict right-angled order according to a uniform plan. These houses still unmistakably characterise the urban features of the town today with its bare, almost monumental monotony. In the basements of these houses, otherwise containing rented apartments, there were as now besides shops the numerous above-mentioned little handworkers' businesses. The craftsmen often worked in the attics, in rooms close to the gables due to the need for light. It is therefore likely that Achille Ditesheim installed his first workshop in the attic of Rue 1er Mars 13, the five gabled windows favourably facing north, endowing it with sufficient light. The workshop could have looked somewhat like the watchmaker's workshop shown on an engraving from the year 1883 (Illus. p. 231 from "Der Mensch und die Zeit in der Schweiz", La Chaux-de-Fonds 1991).

From the La Chaux-de-Fonds address directory, we can see that the location of the company frequently changed in the first few years, as the attic of the house in Rue 1er Mars 13 was hardly adequate for the needs of an expanding watchmaker's workshop, being unsuitable for extension due to its densely built surroundings. In 1892, the company moved to Rue Demoiselle 47; in 1900 it was Rue de la Serre 61. The company did not reach its final location until the construction of a modern factory at Rue du Parc 117 in 1905.

The young watchmaking factory grew fast: in 1890 it employed 30 workmen, and in 1897 their number had increased to 80. For comparison: in 1889, LeCoultre was the biggest employer in the neighbouring Canton of Vaud with 100 employees,

6
A portrait of Achille Ditesheim.

and in the same year, Audemars Piguet was the third biggest with 10 workmen. We can thus see that the Ditesheim company was obviously able to establish and prove itself in spite of the quantitatively high and qualitatively superior competition of La Chaux-de-Fonds.

We know very little today about the company programme in those first years. An old company brochure from the year 1948 shows two pages from the cash-book dating from the company foundation year of 1881. There, the assembly of 198 watch movements of all kinds are indicated with their prices, and the annual balance sheet is shown, being balanced at around SFr. 40,000. Today, these largely decorative pages of the 50-year-old brochure have an unexpectedly and unintentionally high documentary value, as nothing else has been preserved from this time. They also lead us to the conclusion that Achille Ditesheim's workshop at first assembled pocket watches, i.e. put together and finished complete watches from individually bought-in parts. At this time, the Ditesheim brothers were already active as inventors, as shown by their first Swiss patent of 25.10.1894 for a special barrel click.

Thanks to various advertisements in the Swiss Trade Gazette, in which trade names and trademarks were entered, we know that young companies used to have their trademarks registered. In this connection, the company's work programme is frequently described. In 1892, we find "Specialité de mouvements en tous genres". In the time between 1893 and 1896 it grew to include "Boîtes, cuvettes, cadrans, mouvements, étuis, emballages de montres, de ressorts, et autres fournitures d'horlogerie". In the first 11 years of his career, Achille Ditesheim seems to have restricted himself to the assembly of watches and the finishing of movements. From 1893, his company also traded with single parts, such as watch cases, dustcovers, dials, movements, cases and packing material for watches, tension springs, etc., which he bought not only for his own assembly, but also to sell en détail to others. In 1896, the articles dealt with and advertised in the trade journal changed, from then on to remain the same: we find only "Montres, parties de montres" and sometimes additionally "étuis et leurs emballages", i.e. watches and their components as well as their packaging. From about 1896, the company seems to have again given up retail trading with watch components made outside and chiefly sold complete watches – probably assembled in the workshop.

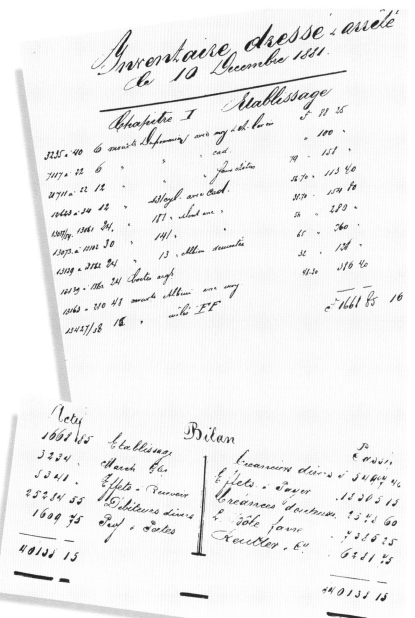

7
Extract from an accounts book dated 1881 (taken from a 1948 catalogue).

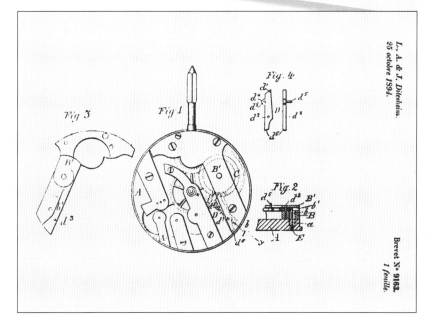

8
Swiss Patent No. 9163 dated 25th October 1894, granted to L.A. & I. Ditesheim for a barrel ratchet mechanism for pocket watches.

There is something else that we are able to learn from the advertisements in the Trade Gazette: the enterprise was now doing business under the name "L., A & I. Ditesheim, fabricants". The reason for this is that, in addition to Achille, the brothers Léopold, who had first gone to Brussels as a watchmaker, and Isidore, who at the age of 19 had just ended his watchmaking apprenticeship, had entered the company in 1886. This was subsequently called after the initials of the three brothers' first names: "L., A. & I. Ditesheim". Their father Samuel died in 1887; thus he was still able to experience the establishment of the watch factory by Achille and the merger effected by the three brothers. It seems that Léopold stayed in Brussels and managed the first branch of the company there, at the same time functioning as sales manager. Soon after, a second branch was opened in Paris. Its manager was Salvator Lévy, who may have been a relative from the family into which Achille, Isaac and Léopold had married.

A third fact may be gleaned from these fertile and informative advertisements: i.e. the great number of trade names under which the Ditesheims at first marketed their watches. There were, on the one hand, such flowery names as "Ultra", "Apogée", "Record", "Talma", "Noblesse", "Salud", "Negus", "Bonne", "Belgravia", "Sûreté", "Mistral" (whereby we do not know whether this refers to the Provençal poet Frédéric Mistral or the cold fall wind from the same region), and "Tanit" (the chief and mother goddess in the ancient North African Carthage).

There were also more functional names like "Chemin-de-Fer International", "Chemin-de-Fer Franco-Belge", "Chemins-de-Fer Réunis", "Chemins-de-Fer Fédéral Suisse" or "Chronomètre du Train". These last-mentioned names point to the fact that the Ditesheim brothers were also active in the national and international market of railway employees' pocket watches, technically high-quality and accurate timepieces which were especially popular in the USA. In an advertisement dated 2nd May 1903, we already find the designation "Movado" as a trade mark; two years later this was to become the general name of the company. Perhaps they wanted to test the name as to its appeal and to see whether it would make a lasting impression.

But let us now go back to the "beginnings". We were already unable to answer the question on what had caused the innkeeper's family to leave Hegenheim in Alsace for Switzerland, and finally for the next generations to concentrate so exclusively on watchmaking. We are equally unable to answer the question as to what induced the barely nineteen-year-old Achille Ditesheim, who could not long before have completed his training, to open a factory or établissage workshop with 6 dependent employees rather than, as others would normally have done, enter service with a master craftsman open a small workshop for repairs or the production of single components. Maybe it lay in Achille's character to choose a more risky path than the majority of his professional colleagues. He certainly received support from the family, whose closeness has already been demonstrated and is shown in the later involvement of the three brothers in the factory as well as by the constant co-operation between them. They helped to carry Achille's initiative and share the risks involved. Moreover, the busy and almost hectic establishment of many similar factories in the nearby surroundings at this time – also in the same region and branch – must have been a challenge, as a visible expression of the cultural age of the "Gründerzeit", i.e. the Foundation Era.

10
Early mantel clocks.

To mention only a few of the most important names: in 1848, Louis Brandt founded the later Omega company in La Chaux-de-Fonds, the Braunschweig family established the Election company; Constantin Girard-Perregaux opened his company in 1856, followed by Léon Breitling in 1884, Paul Ditisheim (Solvil) in 1892, as well as Marc and Emanuel Didisheim (Marvin) in 1894. In the neighbouring Le Locle, there were Ulysse Nardin in 1846, Charles-Frédéric Tissot in 1853, Georges Favre-Jacot (Zenith) in 1865 and George Ducommum (Doxa) in 1889. In St Imier, Ernest Francillon founded Longines in 1867 and J.F. Jeanneret opened Excelsior Park in 1866; in Biel, the Heuer company was opened in 1860 by Eduard Heuer and Bulova was founded in 1885; in Le Brassus, Audemars Piguet was established in 1875 and Urs Schild opened the Eterna company in Grenchen in 1856.

It has already been mentioned that the advertisements for the registration of trademarks in the Swiss Trade Gazette gave an approximate idea of the initial finishing and dispatch programme of the Ditesheim company, which at first comprised the assembly of watches as well as retail trading with these and with single watch components and packaging. From about 1896, the sales of components, i.e. pure trading, decreased. The news that electricity was introduced to production lines in 1897 and at the same time machines for the grinding of jewels and pallet stones reveals that, with the increasing concentration on finished watches, the work of assembly, i.e. the putting together and finishing of watches from bought-in components, diminished. The production of complete watches in the factory, including also the production of most components, was extended. This is also shown by the rapid growth of the company to 80 employees in 1897; workshops doing purely établissage work were rarely so big. The distribution of these watches via branches, representatives and licence-holders abroad was also organised at a very early stage, as shown by Léopold's work in the Brussels branch, already mentioned above. Even before 1900 there was, besides the branch in Paris, a licence-holder for France and Algeria, with Mr E. Dutran and Mr Laurent. Together with this distribution in French-speaking countries came the added by-product of advertising.

The watch programme comprised the usual gentlemen's and ladies' pocket watches as well as pendant watches. "Specialité de mouvements en tous genres" reads the copy in an 1892 advertise-ment: "specialized in watch movements of all types". It is thereby clear that the Ditesheims' main emphasis was on the simple but accurate watch rather than on the more complicated versions. We surmise this, on the one hand, from the above-mentioned commitment to railway employees' pocket watches. Moreover, we see that already before 1900 the Ditesheims had taken part in chronometer trials at Neuchâtel Observatory, also submitting pocket watches to the official Swiss watch testing office in St Imier so that they could sell these with chronometer certificates. This important aspect of the Ditesheims' output will be discussed later.

In 1901, two particularly flat calibres with lever escapement came on the market, one with 11 lignes, thus for small wrist watches, the other with 19 lignes (43 mm movement diameter) for normal gentlemen's pocket watches. At this time, precision watches in wooden cases (similar to precision military watches), so-called "shop-window" display watches, as well as small table clocks with decorated metal cases made up the production programme. Moreover, the company began to specialise relatively early in wrist watches, particularly in small calibres for ladies' wrist watches. In 1895, L., A. & I. Ditesheim were said to have already offered ladies' wrist watches with cylinder escapement.

In the above-mentioned company brochure of 1948, there is a pocket watch movement with lever escapement. It is shown movement side up,

11
Movement of an 1899 pocket watch (taken from a 1948 catalogue).

with the remark that the Ditesheim brothers had begun participating with this movement in the chronometer trials at Neuchâtel Observatory in 1899. If you look carefully, you can see that it is signed "Apogée", i.e. the same trademark that L., A. & I. Ditesheim patented in 1896. This move-

12 a, b
An 18 ct engraved gold hunter cased Movado pocket watch with minute repeating, calendar and chronograph dating from 1905. The enamel dial has a small seconds hand and phases of the moon at 6, a small date chapter ring at 12, apertures for the day of the week and month and is signed, "Movado Sûreté". The gilded movement has a lever escapement and is a Le Phare Calibre LC. The chronograph is operated by the push piece at 2 and the repeating by the push piece at 6.

ment does not yet show the really typical design that was to make the Movado calibre quite unmistakable. It is a normal Swiss movement with bridges and cocks and a separate barrel bridge, as seen in the pocket watch movements of other Swiss watch manufacturers, however with slightly different cock forms. This is probably the oldest surviving Ditesheim calibre made from a bought-in high quality ébauche, which was completed and finely adjusted in the Ditesheim workshops.

At this time, the company also occupied itself with the development of a lever escapement for ladies' wrist watches, whose miniaturisation was certainly made easier by the machines bought in 1897 for the grinding of jewels. At the turn of the century, ladies' wrist watches were generally regarded more as pieces of jewellery than as accurate timekeepers. Thus, more was invested in their appearance than in good rates. It is for this reason that the company stuck so long to the less accurate cylinder escapement for this category of watch. The Ditesheim brothers, however, were among the first to aim also at good performances for ladies' wrist watches. The first, concrete and measurable results of this development, to which we will return again later, come from the time around 1910.

Large international world exhibitions, which had taken place at irregular intervals of several years since 1851, were important events on the world calendar, particularly in the second half of the 19th century. They may be compared to our large industrial fairs today, although they far exceeded these in the scale of the exhibits, with each nation endeavouring to show its greatest technical achievements. The first World Exhibition of 1851 in London produced as a lasting feat the famous Crystal Palace of Paxton, the first large architectural wonder made of steel and glass. Watchmaking was also an important part of these world exhibitions, corresponding to its high position in society at that time. Remember, for example, the shock experienced by the Swiss delegation on being confronted with the overwhelming presence of the American watch industry at the World Exhibition in Philadelphia in 1876. At the Paris Exhibitions of 1878 and 1900, highly complicated pocket watches from the Leroy company drew much attention, also winning prizes. Internationally reputed juries assessed the watches on display, awarding prizes, medals and diplomas, which seem to have had high prestige and advertising value.

These prizes and medals were used intensively for the watchmaking companies for advertising purposes, also by the Ditesheim brothers. They had their first success in Paris in 1900, winning a silver medal. At the World Exhibition in Liège in 1905, it was already a gold medal and an honorary diploma. In Brussels, Movado was awarded the highest prize in 1910, the Grand Prix. Unfortunately we do not know for which watchmaking achievements these prizes were awarded.

The year 1905 represented such a decisive point in the history of our company that it merits a new chapter.

13
Small, engine-turned and enamelled lady's pocket watch in an 18 ct gold case with a very broad bezel. The caseband is set with rubies, diameter 30 mm. The small gilded dial, without seconds, is signed "Chiswell Hnos". Reference No. 5425, nickel silver 11′′′, movement, Calibre 580, 15 jewels, 4 adjustments. Made in 1910.

14 a, b
18 ct gold hunter by Movado with minute repeating from 1908. Silvered dial with small seconds. Lever escapement Le Phare Calibre 102 with repeating release slide in the caseband.

The birth of the MOVADO brand and its first years up to 1910

15
View of the Movado factory
in 1910.

1905 brought our company three innovations that were to affect its future. The first was the construction of a modern factory at a cleverly chosen location directly on the outskirts of La Chaux-de-Fonds on one of the longitudinal main streets, Rue du Parc. With its house number 117, the factory was still near town, but in a district that was not yet densely and completely built up, so that it was possible to buy large plots of land suitable for any later expansion. This factory, planned for over 150 employees, was apparently very modern, including safety measures to an extent that was unusual in those days. For example, the junctions of the belted transmissions, which at that time normally ran freely and openly through the room over the heads of the workers, were specially insulated, allowing the transmissions to run without vibrations or noise. The main part of the mechanical equipment came from the USA. There were moreover good hygienic conditions and the common rooms were light and spacious, not a matter of course at the turn of the century. Separate from the main building, a technical office was constructed for the development of new calibres under scientific conditions, in which there was a department devoted to precision timing for virtually the entire production series. This information comes from an article in the "Revue Internationale de l'Horlorgerie et des branches annexes" of 1906, written on the occasion of the factory's opening. It is one of the quite rare publications on the activities of the Ditesheim family, who were rather publicity shy in any sphere but watches.

This factory was extended by new buildings twice, first in 1917 and later in 1948, in order to house the growing number of staff, which with 300 employees in 1948 had almost doubled vis-à-vis the construction year of 1905, thus making the Movado company one of the biggest enterprises in the Canton of Neuchâtel. It is still standing today and following reconstruction is used as an office building.

The second innovation in 1905 was the arrival in the company of the fourth Ditesheim brother: Isaac, Léopold's twin brother. Isaac was an engraver who ran an engraving workshop in La Chaux-de-Fonds together with the fifth brother Aron, the engine turner. According to contemporary assessment, their company was one of the biggest of its kind in Switzerland. Perhaps Isaac had taken over and continued to run the workshop of his eldest brother Abraham, who died

16
View of the Movado factory
in 1955.

when Isaac was 16 years old. In the 1900 address directory, the output of this workshop is described as follows:

"I. Ditesheim et Frère, Décoration des Boîtes et Cuvettes or, Finissage de Boîtes, ... Spécialité de ramolayés et façons ramolayés. Monogrammes rapportés. Décors à la machine, riches et ordinaires."

Translated, the specialities listed were: Ramolayé, or embossed ornamentation, a decorative technique very similar to chasing, which was used on watch cases; monogram made to order; and simple to intricate machined decoration, of which engine turning was probably the most important.

After these previous successful activities, Isaac was able to afford to invest some capital in the company on joining it, which was certainly welcome in view of the high expenses caused by the new factory. With Isaac's entry into the company, its name was modified to "L., A. I. Ditisheim & Frère", whereby Isaac is represented by the additional "Frère" (brother).

The third innovation in 1905 was the introduction of a new company name, i.e. Movado. We can see how carefully and far-sightedly this new company name was launched from the fact that it had been registered two years before as a trademark, and was still used for several years – at least until 1911 – together with the old company name. Thus at this time the signatures were "Fabrique Movado – L. A. I. Ditesheim & Frère."

The new company name "Movado" comes from Esperanto, the artificial world language that was invented in 1887 by the Warsaw ophthalmologist Ludwig Zamenhof. Movado means "always in motion". Probably the Esperanto circle, which was very lively at this time in La Chaux-de-Fonds, contributed to the Ditesheims' choice of a name from this new artificial language, thus expressing their belief in cosmopolitan attitudes. The company trademark was an image of an uplifted hand holding on its palm an open pocket watch. Very much later, (registered as a trademark in the USA in 1958), the "M" was added over the flat "V": ⋀. This mostly appeared on the dials, while the hand sign was imprinted on the inner side of the case back.

17

The lettering used by
Movado, showing changes
of style over a period of
some decades.

18

A metal plaque given to visitors as a memento of their
visit to the factory.

19

The hand symbol on a large
enamelled metal plate.

20
Travelling alarm clock with
folding support stand. The
case is leather covered and
the midnight blue, enamelled
and engraved dial, with a
small alarm setting chapter
ring at 12, is signed "Alarm
8 Days". Made in 1915,
Reference No. 20385, movement with lever escapement
and 15 jewels.
(Private collection).

21 a, b
Folding 8-day travelling clock
with alarm, which can be
propped up in its leather-
covered case. Gilded
dial, with a small alarm-
setting chapter ring at 12,
inscribed "Movado 8 Days".
Made in 1910, Reference
No. 120634, gilded,
specially shaped movement
with lever escapement,
signed "Movado Fy",
15 jewels, 4 adjustments.

Early patents 1902 – 1912; A summary of the patents up to 1969

In the early years of the present century a series of new inventions and patents commenced. At first they were ascribed to Achille or the Ditesheim brothers, later to the Movado company.

The first two, following the one already mentioned in 1894, are from 1899 and 1902. They are not directly associated with watches but with improvements to an engine-turning machine and the design of a machine for positioning a particular form of decoration called "mille feuilles" ("thousand layers"), which was previously a tedious job carried out by hand.

The next patent, No. 27 690 dating from 17th January 1903, covers a simple date-indication mechanism. This is notable for requiring very little space and is assembled flat onto a plate (see Appendix).

Two further patents were granted on 9th June 1903, both in Achille Ditesheim's name. The first, No. 28 678, is for a winding and hand-setting mechanism for pocket watches. Achille also received an American patent for this, No. 750 619, dated 26th January 1904 (see Appendix).

The second of the two patents dated 9th June 1903, No. 28 679, concerns an improved method of fitting the dial to the movement. This is without dial pillars, a so-called rim-fitting dial (see Appendix).

The next patent, No. 34 976 of lst December 1905, was issued, as all the subsequent ones, to "Fabrique Movado L.A.I. Ditesheim & Frère". It is for a special form of balance spring stud and balance cock which is particularly suitable for thin movement calibres. The stud fixing is hardly thicker than the balance cock itself. This invention takes into account the start of a trend towards ever flatter pocket watches.

A further patent, No. 37 776, was granted on 29th October 1906 for a special form of fine adjustment of the index, particularly for precision watches and chronometers, where the effective length of the balance spring must be adjusted with great precision. Movado was still using this system of fine adjustment for its observatory chronometers in 1930 (see Appendix).

22
Swiss Patent No. 37 776, dated 29th October 1906, for a special fine adjustment regulator index.

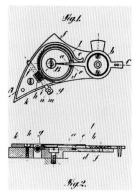

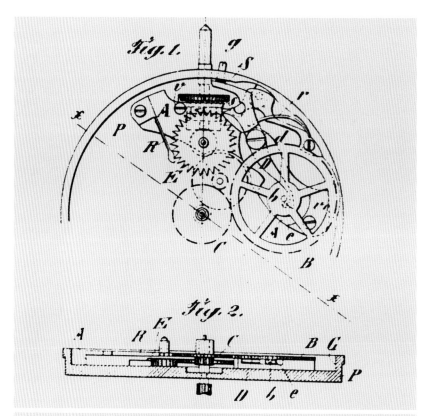

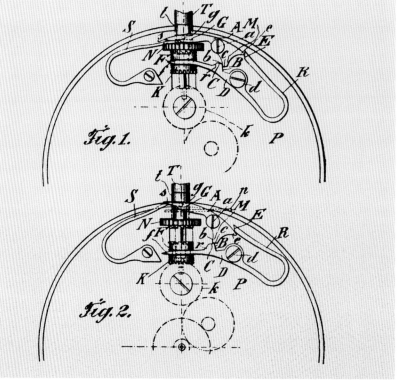

Then followed a patent for a centrally decorated pocket watch dial on 24th September 1907 and two patents (12th February 1908 and 12th October 1908), which were already concerned with the form and attachment of wrist watches and their dials.

With Patent No. 45 161 of 13th November 1908, there commenced a series of five inventions (No. 45 162, 21st November 1908, No. 45 667, 19th December 1908, No. 47 357, 29th May 1909, and No. 51 670, 27th April 1910) for a special form of simplified pocket watch case. These were two-part cases without a fixed middle and caseband. The aim was to achieve particularly flat watches, a style which had come into fashion at that time. It appears that this form of case was predominantly destined for the American market, where access to and removal of the movement from the dial side was usual. This was a continuation of the old English tradition dating from the time of key-wound watches, which was well-known and standard for USA products. However, in Central Europe, including Switzerland, this type of case was unusual (see Appendix).

Patent No. 60 360 followed on 7th July 1912 and described the construction of the Polyplan (see Appendix), which will be discussed in detail later on. In addition to the Swiss one, Movado also received a German Reichspatent, No. 257 360, and an American patent, No. 765 807, for the Polyplan.

The development of technical refinements and inventions and the award of patents is always a sure indication of the advancing creative and innovative powers of an industrial undertaking. In this instance one can only praise the Ditesheim brothers: the 18 patents that we know to have been granted in the decade from 1902 – 1912 are a notable achievement.

The year 1912 represented a decisive point with respect to patents, for thereafter an eleven-year pause occurred. Possibly the First World War and the subsequent post-war years were responsible for this. However, in 1923 a continual stream of inventions and patents began and continued into the 1960's. We can only touch on them here since an intensive study would fill a book in itself.

We will now take a brief look at the patent activities of Movado during the subsequent decades. The frequency of patent awards varies from year to year. The most fruitful year was 1965 with six patents being granted; four relating to mechanical watches and two to electronic watches. Then comes 1930 with five patents granted; perhaps here the beginnings of the world depression, with stagnating sales and the resulting curtailment of production, gave the company the leisure time necessary for research. It is conspicuous, however, that there are no patents between 1931 and 1933.

Comparing the frequency of patents awarded within individual decades, 1900 – 1910, was a highly productive decade with 16 patents, surpassed only by the decades 1950 – 1960 with 19 patents, and 1960 – 1970 with 23 patents.

Besides these numerous practical improvements in Movado's watches there were also a few experiments of a curious and trendy nature. For example: a watch bracelet with a garland of flowers and little bows in 1908; a cuff link with an integrated watch in 1909; a shoulder bag with built-in Ermeto in 1930; a cigarette lighter with a watch; a table clock wound by pulling it back and forth in its frame and, in 1937, a watch with interchangeable decorative plates in the case back. Patent applications for these "playthings" had ceased by the close of the 1930's.

By about 1960 the inventions and patents for electric and electromagnetic drives and control units commenced. They became predominant from 1965 onwards, with Movado jumping onto the "electronic bandwagon" by committing itself to much research in this field.

In all, we can account for the significant figure of at least 98 patents in the period 1900 – 1969. Therefore an average of one and a half patents per year were granted over a span of more than half a century. This record represents an astonishingly constant innovative performance by Movado throughout an extended period of time.

A list of the patents up until 1969 appears in the appendix. 1969 was also the year that the company lost its independence. A selection of the patents between 1900 – 1912 is described in greater detail and many more are at least documented with an illustration. The laborious task of collating these patents was mainly undertaken by Herr Gerhard Streitberger from Munich and I gratefully acknowledge his contribution.

25 a, b, c
Silver Movado double dial pocket watch made in 1915. The normal 12-hour dial with small seconds is signed "Movado Sûreté Suiza". The dial at the back for counting is signed "Contador Movado". The outer black chapter ring has 100 divisions and the inner red chapter ring is divided for 1,000. Each chapter ring has its own hand. Counting is actuated by a push button in the winding crown and return to zero by means of a small slide. The 18¼''', gilded movement with lever escapement is a modified Calibre 821 with 15 jewels. The watch was presumably used in Argentina to count cattle in a field.

26 a, b
A flat, decorated gold Movado pocket watch. The back has a painted enamel scene, showing Napoleon Bonaparte in a garden setting. The two-coloured silvered dial has an engine-turned panel in the centre, no seconds and is signed "Movado Sûreté". Reference No. 17564 5420. Nickel-plated 11''', movement with lever escapement, Calibre 580, with patented attachment of the balance spring stud (engraved with Patent No. 34976). 15 jewels, 4 adjustments. Made in 1915.

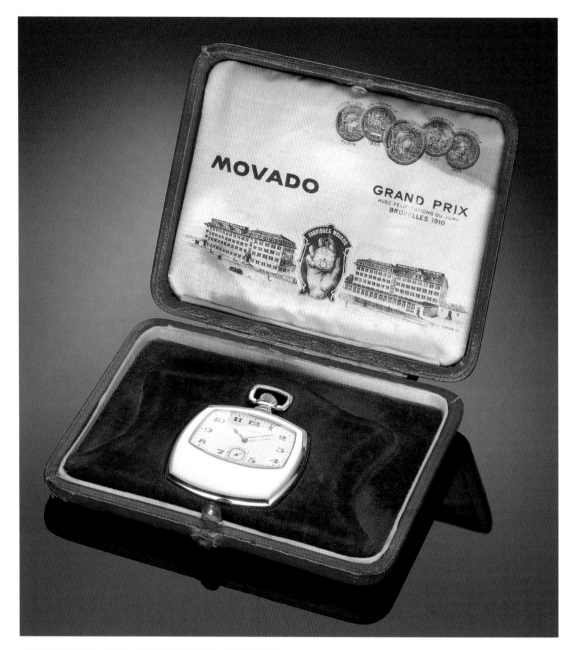

27 a, b
Movado cushion-shaped pocket watch with an 18 ct gold case. The silvered dial with small seconds, signed "Chronomètre Movado", is placed above the centre of the watch. This is achieved by placing the nickel-plated 11''', Calibre 600 movement eccentrically in a larger plate. The movement, signed Movado, has a lever escapement, 15 jewels, 4 adjustments, a simple regulator index and is decorated with "filets". Reference No. 611969 5245, made in 1915.

29 a, b, c
Small, decorated gold hunter with enamel paintings on front and back covers. Diameter 30.4 mm. White enamel dial with small seconds and signed "Chronomètre à Ancre". Reference No. 583828. Nickel-plated Calibre 580 movement with lever escapement, decorated with "filets", 3 gold chatons secured by screws, simple regulator index. (Private collection).

28 a, b
Movado pocket watch in an 18 ct gold case, made in 1915. Silvered dial with engine turning, no seconds, signed "Movado Sûreté". Reference No. 17022 5429, diameter 45 mm. Nickel-plated 11‴, Calibre 580 movement with lever escapement set into a larger plate. Signed Movado Watch Fy, 15 jewels, 4 adjustments. 3 gold chatons secured by screws. Patent No. 34 976 on the balance cock refers to the special fixing of the balance spring stud, simple regulator index.

30 a, b
Movado pocket watch in
18 ct gold case with a broad
bezel. Reference No. 7110.
Gold-coloured dial with small
seconds. Nickel-plated
16¼''', movement, Calibre
620, with lever escapement,
17 jewels, 3 gold chatons
with screws, fine adjustment
regulator according to Patent
No. 37 776 (engraved on
the balance cock). Made in
1910.

31 a, b, c
Movado 18 ct gold hunter
with a large diamond set in
the decorated front cover.
Case diameter 33 mm.
Enamel dial with small
seconds signed Movado.
Reference No. 5118 309
588309. 11''', Calibre 580
nickel-plated movement with

"filets" decoration,
15 jewels, 3 gold chatons
with screws, signed
"Movado Sûreté", 4 adjustments, uncut balance with
screws, simple regulator
index. Made in 1915.

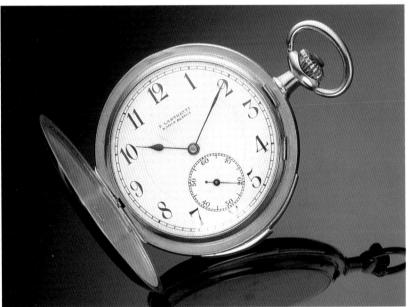

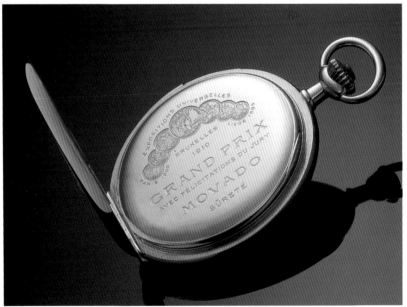

32 a, b, c
Movado 14 ct gold hunter
with quarter repeating, made
in 1910. Silvered dial with
small seconds, signed
"F. Lastrelli Bahia-Blanca".
Nickel-plated movement with
lever escapement, Le Phare
Calibre 105, 18 jewels.

The "Polyplan"

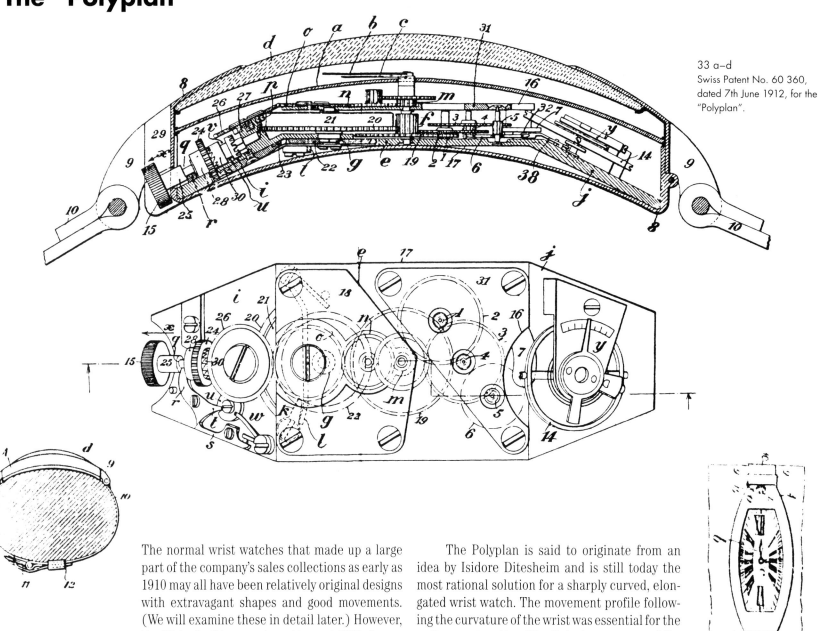

33 a–d
Swiss Patent No. 60 360,
dated 7th June 1912, for the
"Polyplan".

The normal wrist watches that made up a large part of the company's sales collections as early as 1910 may all have been relatively original designs with extravagant shapes and good movements. (We will examine these in detail later.) However, the "Polyplan" towers above the rest. With its unconventional construction it is considered to be unique and is treated in much the same way as the "Duoplan" or "Reverso" by Jaeger LeCoultre in most books about the history of the Swiss watch (e.g. Plate 154 in "Histoire et Technique de la Montre Suisse" by E. Jaquet & A. Chapuis, Basel/Olten 1945).

The Polyplan is said to originate from an idea by Isidore Ditesheim and is still today the most rational solution for a sharply curved, elongated wrist watch. The movement profile following the curvature of the wrist was essential for the entire construction. The Polyplan was one of the first "form" movements developed specifically for curved wrist watches, a style which was very fashionable in the years before the First World War.

The designated task was to equip a wrist watch of a very elongated form with a movement that followed the curve. The possibility of making a conventional straight-line (or baguette) movement or a round movement was limited by the curve of the case and the narrow (approximately 15.2 mm wide) available space. The balance, in particular, with a maximum diameter of 5 mm, and the escapement part of the movement, were so constricted that only an inferior rate would be achieved. At this time the mass production of a miniaturised version of the lever escapement was in its infancy. In addition, the performance of tiny movements with very small balances is further limited by the proportionally greater mass of the lever, compared with larger calibres. The apparent goal for such a wrist watch was to have a rate performance as good as could be expected from an 11′′′ gentleman's wrist watch with a 9 mm diameter balance. Even movements with what later became the common layout, having a level train plus partially angled bridges and cocks, were not curved or large enough to accommodate a large balance wheel.

The unique and permanent solution to this problem by the Movado engineers was to angle both ends of the movement plate downwards by 25°. This arranged the movement on three planes, such that it could follow the curve of the case. The name "Polyplan", is derived from these planes: as polygon means many angles, "Polyplan" signifies many planes. The balance is placed on one of the angled planes, beneath a cock placed at right angles. The connection with the train on the middle portion of the plate is achieved by means of an appropriately angled portion of the pallet lever. The other angled part of the plate carries the winding-stem and crown, situated at 12, as in a pocket watch, and the intermediate hand-setting wheel. The latter engages the minute wheel, positioned on the centre portion of the plate, at an angle of 25°. The conventional train and barrel are also positioned on the centre portion of the plate.

The second unconventional feature of this movement, Calibre 400, is that it is constructed on the so-called bagnolet principle, in which the wheel side of the movement is under the dial, rather than as usual beneath the back of the case. This allows a somewhat larger balance to be used and the whole watch becomes flatter. An observer, on opening the back, will find a rather, uninteresting, practically unbroken, plate surface. The escapement, balance and other parts are concealed beneath the dial. The latter is not fixed to the movement. It is clamped between the upper and lower portions of the case. Movado was grant-

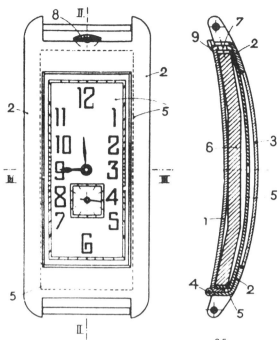

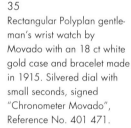

34
Patent No. 130 940, dated 16th March 1929, for an alternative dial fixing for the Polyplan.

35
Rectangular Polyplan gentleman's wrist watch by Movado with an 18 ct white gold case and bracelet made in 1915. Silvered dial with small seconds, signed "Chronometer Movado", Reference No. 401 471.

36 a–f
Special nickel-plated, angled form movement Calibre 400, Movement No. 401411, signed "Movado Factories", 15 jewels, 3 gold chatons with screws, special regulator index to be operated from the back (see Illus. 35).

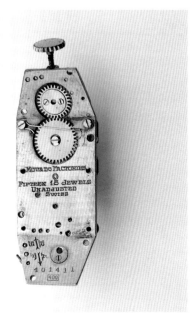

37
Two Polyplan wrist watches in different tonneau-shaped cases, one silver and the other 18 ct gold, made in 1915. The movements are Calibre 400.

ed a further patent on the 16th March 1929 to protect a rectangular Polyplan construction, whereby the movement is inserted into a dial with all four sides bent down, which is then placed in a correspondingly angled case back.

The bagnolet calibre, also known as a reversed calibre, was developed for pocket watches around 1800 by the renowned watchmaker Philippe-Samuel Meylan from the Joux Valley. Very flat movements, such as were fashionable during the first half of the 19th century, may be constructed on this principle. Meylan was one of the first watchmakers to fit these flat bagnolet movements into coin watches.

The Polyplan construction was protected by patents in the major Western countries; in Switzerland by Patent No. 60 360, in Germany by Deutsche Reichspatent No. 257 360 dated 11th June 1912, and in the USA by US Patent No. 765 807. The movements were manufactured in the original form between 1912 and 1917. The patent dated 16th March 1929 with a modified dial fixing gives the impression that at this date Polyplan movements were still (perhaps, again) being made, or possibly completed in small series from old stock.

Polyplan movements, which were generally cased in very slim elongated oval, or extended tonneau-shaped gold cases, were not for everyday use. They were too expensive for that, presumably due to the considerable effort required to produce them. This probably prevented a large quantity being made. With Movado it is possible to establish the magnitude of a production run for a particular calibre from the movement numbers. The highest recorded movement number for Calibre 400 is No. 1471, indicating a total of 1,500 examples. However, there are quite possibly more. Furthermore the era of the Second World War was approaching, bringing quite different problems than elegant and expensive luxury watches. Today, the Polyplan has become a relatively rare, sought after and expensive vintage wrist watch. It is instantly recognisable by its very slim and highly curved case form, as well as the winding crown positioned on the upper short side, above the 12. It exists with a small seconds hand, between the centre and 6, as well as without a seconds hand. The American author and vintage watch dealer Charles Cleves supposed that the Polyplan was the inspiration for the Gruen Curvex in the thirties (Movado – ahead of its time: Horological Times, November 1988). However, the latter was more conventionally designed.

15 years of Movado between 1910 and 1925

39
Two "Polyplan" models.

Two Movado brochures from the decade 1910 – 1920, in which the selection of models has been captured photographically, have survived. They illustrate 704 different wrist watches, of which 80% are ladies' models and only 20% gentlemen's. The difference between the ladies' and gentlemen's models is often not obvious since Movado manufactured many so-called medium-sized watches which in those days could be worn by either sex. We shall encounter this Ditesheim tendency again with the "Ermeto". In addition, these catalogues show 40 pendant and button-hole watches, a few table clocks and 381 pocket watches. Amongst the latter are versions for gentlemen and some small decorative, sometimes extravagantly shaped, ladies' watches. Also shown are a few complicated pocket watches with repeating, chronograph and/or calendar functions in the then conventional executions. Over 1,000 different models!

We do not know with certainty if this represented the whole selection available at this time, but it seems probable, however, due to the large number of styles. In the case of the wrist watches, and particularly the ladies' models, an endless diversity of sheer fantasy is apparent. Very imaginatively shaped watches with a great variety of decoration were available, with modern and sometimes extravagant case designs: round, many-sided, horse-shoe shaped, horizontal and vertical ovals, and fantasy forms. This was no doubt the work of Isaac the engraver, who joined the firm in 1905, and was responsible for the design of both case shapes and decoration.

This wide variety of models does not, of course, give any clue as to the actual quantity and distribution of individual models. Many of the richly decorated ladies' wrist watches were probably produced in series of only 5 – 10 examples. We do not know Movado's annual production figures for this decade. With about 1,000 different models available, 50 of each per year, we arrive at total of 50,000 watches, a remarkable annual production. As a basis for comparison, the annual production during this period was about 2,800

40
Page from a 1915 Movado
model catalogue showing
gentlemen's wrist watches,
including 2 early chrono-
graphs with push button in
the winding crown.

41
Page from a 1911/1912
Movado catalogue showing
the models available at that
time.

42
Early tonneau-shaped 18 ct gold gentleman's wrist watch of half hunter type, made in 1915. Reference No. 548378-5720 with 11¾''', Calibre 530 movement with lever escapement, 15 jewels. Case and movement signed "Movado".

43
Page from a 1911/1912 Movado model catalogue.

44
Page from a 1911/1912
Movado model catalogue.

45
Page from a 1911/1912
Movado model catalogue.

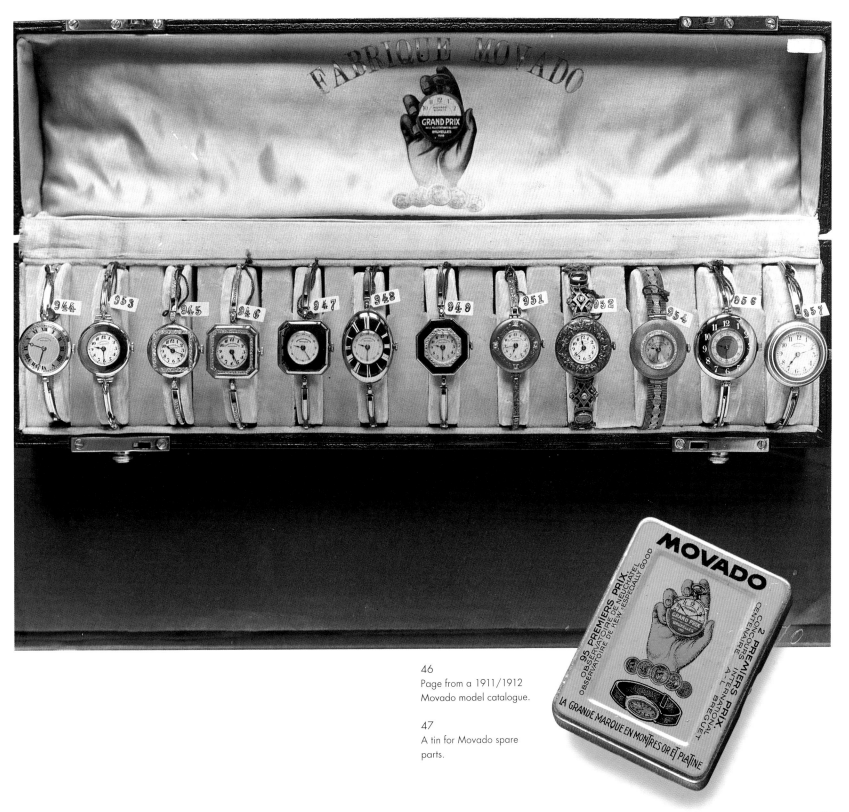

46
Page from a 1911/1912
Movado model catalogue.

47
A tin for Movado spare
parts.

watches for Audemars Piguet, some 26,000 at IWC and approximately 65,000 watches at Omega. Nevertheless, the distribution of the models amongst the watch categories is both interesting and informative.

It is not unusual that the clear majority, some 80% of the range, were ladies' wrist watches, since gentlemen's models are simply plainer and thus require fewer variants. More interesting, and a sign of Movado's early and intensive move towards wrist watches, is the fact that about 65% of the total model range offered consisted of wrist watches, and that these already significantly dominated the pocket watches. Moreover, this was at a time when other renowned Swiss watch manufacturers had either not yet started or were just beginning the production of wrist watches. However, in terms of actual Movado watches produced, the figures mentioned above may not reflect the true picture.

Amongst the gentlemen's wrist watches being produced at this early date are two large single push piece chronographs. The functions of start, stop and return to zero are achieved by a single push piece in the winding crown, the same as in all the very early chronographs from this decade. The dial is arranged with a small seconds dial at 9 and a corresponding dial with a register for 30 minutes at 3. In 1912, Movado was, together

48
An early wrist chronograph with split seconds, signed with the Movado registered brand name Ralco.

with Omega and Ulysse Nardin, among the first Swiss companies to produce wrist chronographs.

A large part of the Movado range was not signed as such. Commercial names such as "Mistral", "Tanit" or "Ralco", as already mentioned in some detail, were also employed. "Ralco" was a

trademark stemming from the initial letters of three further members of the Ditesheim family. They were from the next generation and already active in the business: Roger, Armand and Lucien. Thus, RAL plus Co equals Ralco. This name was predominantly used for inexpensive watches of a lower quality, since such watches might have tarnished the image of the Movado name as a precision watch company. However, some of the most valuable watches were also sold under this name, such as a 1921 split seconds wrist chronograph (a chronograph with an additional, independently controlled, seconds hand) and a minute counter to record up to 30 minutes (see: Gisbert L. Brunner, Armbanduhren, Munich 1990, page 135). Some of the models were signed, in addition to "Movado", with "Rosenberg-Wallach, Joyeros, Lima Peru", presumably the name of the Peruvian or South American representative of Movado. Others were signed "Chronometro Escasany".

It is also noteworthy that more than half of the wrist watches – 436 to be exact – shown in the two brochures are signed "Chronomètre" alongside "Movado". This even applies to the smallest and richly decorated ladies' dress watches and many of these "chronometers" do not even have a seconds hand. However, one must realise that an officially approved definition of the term "chronometer" for wrist watches did not appear in Switzerland until 1926, when it was the first country of all to define the term. (Translator's note: Chronometer in English often refers to the escapement of this name, as used in marine chronometers, but this is not the case here.) One could hardly compare the significance of the term chronometer with what, after the official 1926 definition, became the standard of accuracy, in particular with regard to very small ladies' dress watches without seconds hands.

It may also be observed that many early wrist watch calibres (50 SP, 105, 150 MN, 470) are marked "4 four adjustments", i.e. four things have been attended to. However, it remains unknown if the watch has been checked in two positions, at two different temperatures, or in four different positions. To attain the standard of an officially certified wrist chronometer, after 1926, four adjustments were no longer sufficient. By then 7 adjustments were required, 5 positions and at 2 temperatures. Nevertheless, there are Movado wrist chronometers from the post 1926 period until well into the 1940's with only 4 adjustments. Perhaps the fact that the very small 1905 5 1/2 ligne Calibre and the small form movements 9 M were both equipped with lever escapements was good enough reason, at least within the firm, to designate them as chronometers, which was remarkable enough for this time. Around 1910, it was completely new and unique for wrist watches to be finely adjusted in large numbers to such a degree that they had rates that could be said to at least approach those of a chronometer. The standards for marine chronometers, deck watches and pocket watches, established at least 40 years previously, were not so far removed from the conceptions of Movado that they ran the risk of being deemed untrustworthy. Movado was a pioneer in this field. It should be pointed out that many of the later renowned wrist chronometer manufacturers, such as Omega, Eterna and Zenith, only started to make, rate and submit their watches for official chronometer certification in the thirties and forties. The oldest established firm in this field, Rolex, commenced in 1927. The performance of Movado as a precision watch manufacturer, as confirmed by their efforts and successes in observatory trials, has hardly been recognised and praised. They particularly deserve recognition for their early precision wrist watches.

The movement calibres

Not surprisingly, the brochure illustrations show just the dial side of the watches. This was the chief aspect of interest to the customer of the day. Only the watchmaker was interested in the internal parts; the movement became important to the customer only if the watch was not working well. The watch collectors of today are different; for them the movement is also significant and a brochure only illustrating dials would soon become boring. So it is a stroke of good fortune that a page in one of the two brochures illustrates movements. We must assume, however, that this page of 15 (14) factory illustrations, shown full size, does not represent the complete product line since the small 5 1/2 ligne Calibre of 1905, the 7 1/2 ligne Calibre 50 SP, which must already have been in production at this time, and the small form Calibre 9 M with the larger version, Calibre 11, are all missing.

The 15 movements arc ascribed little flags with three digit numbers (154 – 169). The meaning of these numbers is no longer clear but they are useful to designate the movements. The movement belonging to 159 is missing. In the case of the pocket watch movements, two calibres appear twice in different quality versions. It would appear that Movado had about 18 of their own calibres in production for the period 1910 – 1920. Let us now deal with these movements individually. They are all marked "Movado Fᵞ", the abbreviation standing for "Factory".

The movement illustrations date from about 1915. Seven of the movements illustrated are for pocket watches, one of which (165) is a hunter movement. Of the other eight, four are wrist watch movements (158, 160 – 163), and the remaining three, judging by the positions of the winding crown, balance and fourth wheel, were intended for pendant, button-hole or ladies' watches. With two exceptions, all the movements, both large and small, possess the basic form which characterised almost all Movado movements up until the 1950's. A very characteristically shaped train wheel cock with curved ends and three rounded projections (bulges), which accentuate the escape wheel, fourth wheel and third wheel bearings; a curved half plate for the barrel and ratchet wheel with a semi-circular projection for the centre wheel bearing, and a simple but elegantly curved balance cock. The differences between the movements illustrated lie in the size and the final quality, as we shall see.

The exceptions to these movement layouts are for two wrist watch calibres. Movement 158 is likewise round but has an acutely angled cock form. We shall encounter this movement again in a military watch but in a noticeably higher developed form – the present example does not even have jewelled bearings. The second exception concerns the only non-round movement and is, in fact, the elongated Polyplan (161), where an under-dial view is shown.

Some of the movements may be identified with a high degree of certainty by their calibres. Among the seven pocket watch movements, Nos. 154 and 155, are presumably differently finished versions of Calibre 800 M. No. 154 is with côtes de Genève plate decoration, chatons with 3 screws, black polished winding wheels and a special precision index according to Patent No. 37776 of 1906. No. 155 is somewhat simpler with a so-called damascened finish (popular in the USA) and a simple index. Only two of the pocket watch movements (165 & 166) have factory numbers, 360 021 and 340 018. Amongst the small movements, only the Military Watch Calibre 420 (158) and Calibre 400, Polyplan (161), are identifiable. No. 160 appears to be Calibre 470 and No. 162 the universal wrist watch Calibre 150 MN.

Where indicated, the movements have either 15 or 17 jewels and four, or less commonly five, adjustments. Both of the pocket watch calibres with factory numbers have seven adjustments and 21 jewels. They also have a very high quality finish and are free sprung, i.e. without index. They were clearly destined for chronometer trials and competitions. The train bridge of No. 360 021 carries the words: "Chronomètre de Bord Observatoire de Neuchâtel". This watch was entered in the deck watch category of the 1913 trials, achieved 20th place and was awarded a second prize. In the previous year it had been submitted as a pocket chronometer and was less successful, with 80th place and a third prize. This watch belongs to the first series of observatory movements with a plate diameter of 43 mm (see p. 54 f).

Some of the 14 movements (157, 162, 167, 169) carry a reference to Swiss Patent No. 34976 on the balance cock. This is the 1905 patent, previously mentioned, for a particularly flat form of balance spring stud fixing. Worthy of note is that all movements up to the atypical Movado Calibre 420 (158) were regulated with at least 4 and, at the most, 7 adjustments. Even the smallest pendant watch movements (164 and 169) still had 4 adjustments, although it must be stated that the wording on the pocket watch movements Nos. 155 and 156 is not entirely legible as here the engraving was not black filled. The numerous adjusted movements confirm the statement made earlier that Movado pioneered precision movements very early on, earlier than their competitors, even trying to achieve good rates with small wrist watches, as far as production and series adjustment allowed. Thus, in many cases, the watches could be classed as chronometers.

The year 1914, black and fateful as it was for Europe, brought Movado, for the time being, further fortunate circumstances so that it could almost be compared to the eventful year of 1905. First of all business was so good and the demand for their products so high that the factory in La Chaux-de-Fonds had to be enlarged. Isidore Ditesheim, the inventor of the Polyplan, was appointed to the jury of the Swiss National Exhibition in Berne – a national equivalent to world

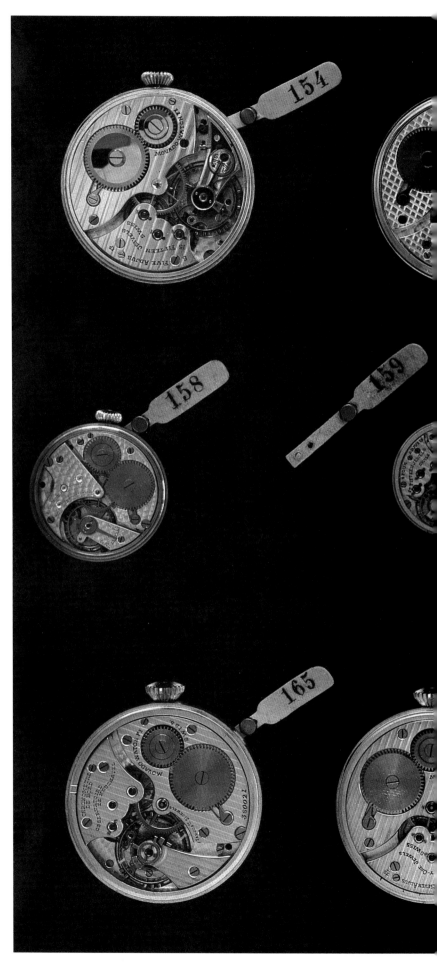

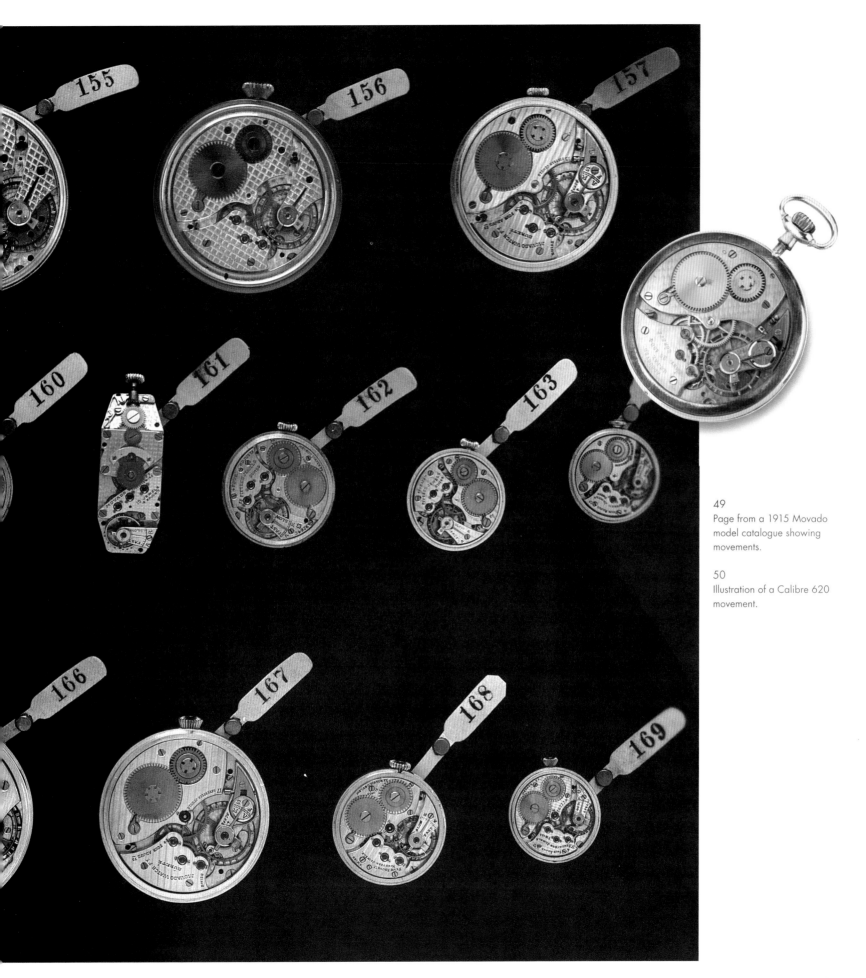

49
Page from a 1915 Movado
model catalogue showing
movements.

50
Illustration of a Calibre 620
movement.

exhibitions. Finally, it was the start of the First World War that led to the development of the military wrist watch, which was to become one of the biggest economic successes. The start of the war, however, froze all business activities temporarily. In spite of its neutrality, Switzerland was cautious and called up all men eligible for military service. This left Movado and all other factories deserted for an uncertain period of time, including Movado's newly completed extension to its factory building.

wearer to read the time from the wrist rapidly, without the awkwardness of opening the protective cover. Wrist chronographs were also used during the Great War. They were of the single push type, with the push piece and winding crown positioned at 9 for better protection and large luminous numerals. With these watches it was possible, on seeing the muzzle flash, to rapidly calculate the range of enemy fire. Protection from moisture, a major problem in the trenches, particularly during lengthy trench warfare, could be improved by means of screwed cases, already in use at that time.

Eugène Jaquet and Alfred Chapuis wrote in 1945, in their great (and so far unique) book about the history and technique of the Swiss Watch, that Movado was the initiator of the military watch with pierced cover. They illustrate

The First World War Military or "Soldier's Watch"

It has been claimed that, in 1880, the German Navy ordered gold wrist watches with gold bracelets from Girard Perregaux in La Chaux-de-Fonds. However, no examples of these watches have survived and it is not known what they looked like. Judging by the use of gold in these watches, it does not seem likely that they were foreseen as practical watches destined for hard military service. The Prussians were not that extravagant! In any case the fact remains that an army ordered a significant number of military watches, which became the forerunner of the later specific-purpose trench or military watch. Also in the South African Boer War, around 1900, wrist watches were said to have been used in large numbers and withstood usage under the hardest conditions.

The question of which company undertook the development of this type of military watch has not yet been thoroughly researched. The earliest known wrist watches specifically for military use were, once again, ordered by Girard Perregaux. They are supposed to have supplied such watches, as before, to the German Navy, this time in 1910. By then, these watches had an important and unmistakable addition in the shape of a pierced metal cover over the dial and glass to protect these fragile components from damage in the event of hard usage. Further characteristics of military watches are: robust construction, not too small, clear dials with good contrast, luminous numerals and hands for telling the time in the dark or poor light of the trenches and a solid case of either steel or silver. It was the first time that wrist watches were undeniably produced for a definite purpose. The pierced cover also enabled the

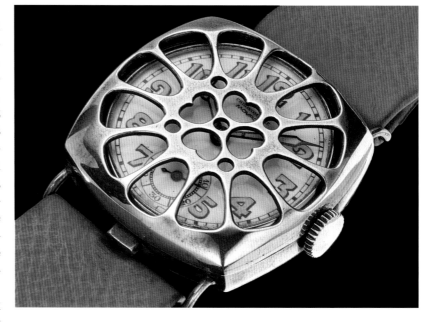

(Plate 154) just the same watch that can be seen here. This model has meanwhile been identified as a development from 1914. Therefore, it appears that Girard Perregaux were somewhat earlier in this field.

Many Swiss watch companies supplied military watches to the armies at war between 1914 and 1918. This was a lucrative business, involving the large quantities that were welcomed and urgently needed by many a business that had suffered losses in sales abroad between 1908 and 1910, following the 1907 New York Stock Exchange crash. Of the subsequently well known military watches (besides Movado and Girard Perregaux, other known brands included Eterna, Omega, IWC, Longines, Ulysse Nardin, Enicar, as well as the American firms Waltham and Inger-

551 a–c
The First World War Soldier's Wrist Watch of 1914. Cushion-shaped silver case with pierced protective cover over the dial, which has luminous numerals, small seconds and is signed "Chronomètre Movado".

Reference No. 24463 452 259. Nickel-plated 13''', Calibre 420 movement with lever escapement, circular spotted decoration (Oeil-de-perdrix), 17 jewels, 4 adjustments, 4 gold chatons with screws, simple regulator index. Movement unsigned.

51

soll) the Movado was clearly the best developed, both aesthetically and technically, even though it was developed in record time in 1914 when the factory was partially closed owing to a lack of workers.

This particular example of a Movado Military Watch has a quality 13‴ (29.3 mm) lever movement with a fine Oeil-de-perdrix decoration in concentric circles on all the bridges and cocks. It was designated Calibre 420. As already mentioned, this calibre differs from the typical and characteristic Movado form, although it is still, without doubt, a Movado. The firm protected this calibre in 1915 under Patent No. 25100. A noticeable characteristic is that the train bridges run into a sharp angle. Quality is apparent with a compensated bimetallic balance, 4 chatons with screws, and 5 adjustments. The latter is surprising for an early wrist watch, the more so since it surpasses the tradition of Chronomètres Movado with only 4 adjustments. This movement also exists as Calibre 480 in Lépine form, i.e. turned through 90° so that the winding crown is at 12 and the small seconds is diametrically opposed at 6. In this form it is suitable for ladies' pocket or pendant watches.

This movement is cased in a tonneau-shaped silver case with an imaginatively pierced protective cover that leaves the large heavy numerals fully exposed. With its comparatively large 13‴ movement – the largest Movado wrist watch movement at that time – the complete watch measures a significant 35 x 35 mm.

Movado's Military Watch (or Soldier's Watch) was developed in record time for the outbreak of the First World War and went into production shortly after the mobilisation order was withdrawn by the Swiss authorities. The enforced quieter interval was soon compensated for by large series of this model, which was probably also the reason that the factory building required enlarging again in 1917. In the same year the firm became a limited company (societé anonyme) with the name changed to "Fabriques Movado".

The total production quantity of the Military Watch can only be estimated. The highest known movement number is 427 800. This means that, at the time this watch was being manufactured, 7,800 movements had been produced. However, although it is unlikely that this was actually the final total, it does point to the fact that, assuming wrist watches with this movement were only produced during the First World War, around 2,000 examples were made per year. At any rate

more than the new edition of 1993, limited to 250 watches.

52
Drawings for the registered design of the Calibre 420 movement used in the Soldier's Watch. Above, the hunter (wrist watch) version, below the open-face version.

53 a, b
Cushion-shaped silver gentleman's wrist watch from 1917. Silvered dial with luminous numerals and small seconds, signed "Movado", Reference No. 427 800. Nickel-plated 13‴, Calibre 420 movement (as in the Soldier's Watch) with lever escapement, signed "Movado Fy", 15 jewels, 4 adjustments, simple regulator index.

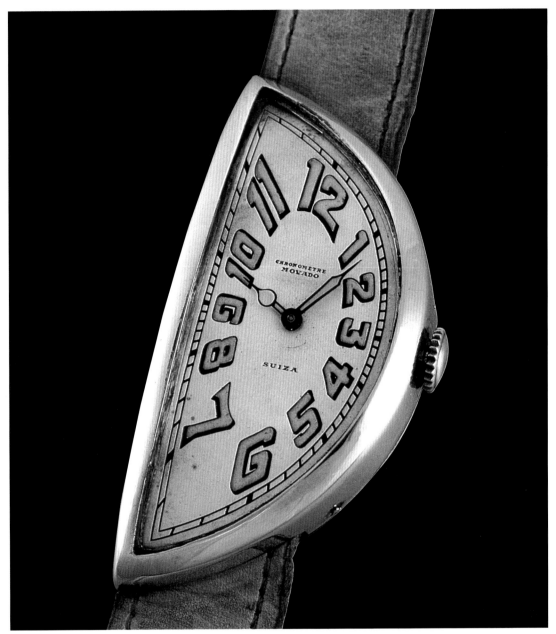

54 a, b
Early gentleman's wrist watch in unusual, half moon-shaped white gold case, made in 1914. Silvered dial with bold luminous numerals, signed "Chronomètre Movado Suiza", no seconds. Nickel-plated 10¼''', Calibre 150 MN movement with lever escapement and "filets" decoration, signed "Movado", 15 jewels, 4 adjustments, 3 gold chatons with screws, simple regulator index.

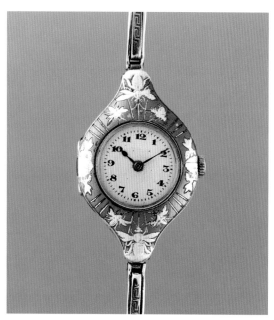

55 a, b c
Early decorated silver lady's wrist watch. This watch was a prize from a Canton of Neuchâtel shooting competition in 1913 (engraved on the back). The unsigned silvered dial is without seconds. Nickel-plated movement with lever escapement signed "Movado Watch Fy, Sûreté", 15 jewels, 4 adjustments, 3 gold chatons with screws, simple regulator index.

56

Two early 18 ct gold ladies' wrist watches. The horseshoe-shaped watch on the left has a silvered dial signed "Chronomètre Movado" and small seconds. Nickel-plated 8¾''', Calibre 780 movement with lever escapement, 15 jewels, 4 adjustments. The watch on the right has a broad bezel set with diamonds, silvered dial without seconds, signed "Movado", Reference No. 9781-933537. Nickel-plated 9½''', Calibre 950 movement with lever escapement, 15 jewels, 3 gold chatons with screws.

57

Early 14 ct gold wrist watch with broad bezel and date indication in an aperture, made in 1915. Unsigned silvered dial without seconds. One of the first wrist watches with a date indication. Nickel-plated 11''', Calibre 580 movement with lever escapement, 15 jewels, 3 gold chatons with screws.

58 a, b

Early cushion-shaped Movado lady's wrist watch in 18 ct gold case, made in 1915. Silvered dial with small seconds signed "Movado", Reference No. 95, Serial No. 785913. Nickel-plated 8¾''', Calibre 780 movement with lever escapement signed "Movado", 15 jewels, 4 adjustments, 3 gold chatons with screws, simple regulator index and "filets" decoration.

60
Early square 18 ct white gold wrist watch, made in 1917. Silvered dial, signed "Movado", without seconds. The case was made in the USA. Nickel-plated 9½''', Calibre 950 movement with lever escapement, 15 jewels.

59 a, b
Early 18 ct gold Movado lady's wrist watch with enamel painting, made in 1915. Slender, tonneau-shaped case with colourful flower painting and movable strap attachments. Cushion-shaped dial set cornerwise, unsigned and without seconds, Reference No. 6009, Serial No. 551 027. Nickel-plated form movement, calibre 550, signed "Movado", with lever escapement, 15 jewels, 4 adjustments, 3 gold chatons with screws, simple regulator index.

61 a, b
Early rectangular 14 ct gold Movado wrist watch, made in 1917. Silvered dial with small seconds, Reference No. 157 255, Serial No. 41543. Nickel-plated 10¼''', round movement, Calibre 150 MN, signed "Movado", with lever escapement, 15 jewels, 4 adjustments, 3 gold chatons with screws, simple regulator index.

63
Horizontal oval silver-cased
gentleman's wrist watch,
made in 1920. Silvered dial
with small seconds, signed
"Chronomètre Movado".
Major axis 46 mm.
Reference No. 2905-1377-
531081. Nickel-plated
11¾''', Calibre 530 move-
ment with lever escape-
ment, 15 jewels, 3 gold chatons
with screws.

◁ 62
Two early silver Movado
wrist watches in extra-
vagantly shaped cases from
1920. Both signed "Chronomètre Movado". One move-
ment with Calibre 530, the
other Calibre 580, both with
lever escapement.

64 a, b
Early 18 ct gold-cased gentleman's wrist watch with
chronograph operated by a
push button in the winding
crown, made in 1920. Black
dial with gold numerals and
hands, calibrated to measure
pulsations, small seconds
and 30-minute counter,
signed "Movado", Reference
No. 147, Serial No.
246560. Movement 13''',
with lever escapement,
column wheel chronograph
mechanism, presumably
Valjoux Calibre 13.

66

Gentleman's wrist watch in striking bi-colour 18 ct gold case, made in 1924. The watch is contained in a round, yellow gold case, which is set in an open, rectangular, white gold frame forming the case lugs for the leather strap. Silvered dial (reconditioned) with Hebrew numerals, small seconds, signed "Movado", Reference No. 9735, Serial No. 957278. Nickel-plated, 9½''', Calibre 950 movement with lever escapement, 15 jewels, 3 gold chatons with screws.

65 a, b

Movado 14 ct gold wrist watch, made in 1920. Silvered dial with small seconds, signed "Chronomètre Movado", Reference No. 49 752. Nickel-plated with "filets" decoration. 9½''', Calibre 950 movement, signed "Movado Fy", with lever escapement, 15 jewels, 4 adjustments, 3 gold chatons with screws, simple regulator index.

67

Two early Movado wrist watches in unusual cases. The watch on the left is in a cushion-shaped 14 ct gold case, made in 1915. On the right, a watch in a silver, horizontal oval case. Signed "Chronomètre Movado". Both watches have nickel-plated, 9½''', Calibre 950 movements with lever escapement. The leather straps are original.

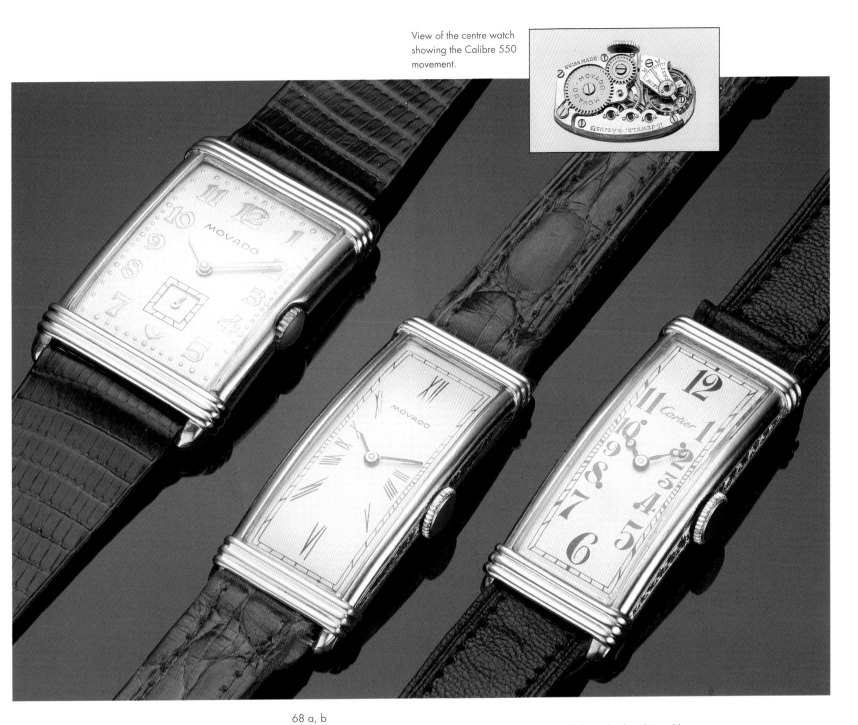

View of the centre watch showing the Calibre 550 movement.

68 a, b

Three rectangular Movado wrist watches, made in 1920. Left, a bi-colour yellow and white gold cased watch, with silvered dial, signed "Movado", small seconds, Reference No. 42, Serial No. 742 625. Round 8¾''', calibre 730 movement with lever escapement, 4 adjustments.

Centre, a very slender yellow-white gold cased watch with silvered dial, fine numerals, signed "Movado", no seconds. Form Calibre 550 movement with 4 adjustments.

Right, a slender white gold cased watch with silvered dial, signed "Cartier", no seconds, Reference No. 6035, Serial No., 554327. Form Calibre 550 movement with 4 adjustments.

69 a, b
Large gentlemen's wrist watch with chrome-plated case, made in 1920. Silvered dial with luminous numerals and hands, small seconds, signed "Movado Non Magnetic", Reference No. 37874 0631648. Nickel-plated, 16¼''', Calibre 680 movement with lever escapement, 15 jewels, 4 adjustments, 3 gold chatons with screws, signed "Movado". View of dial and movement.

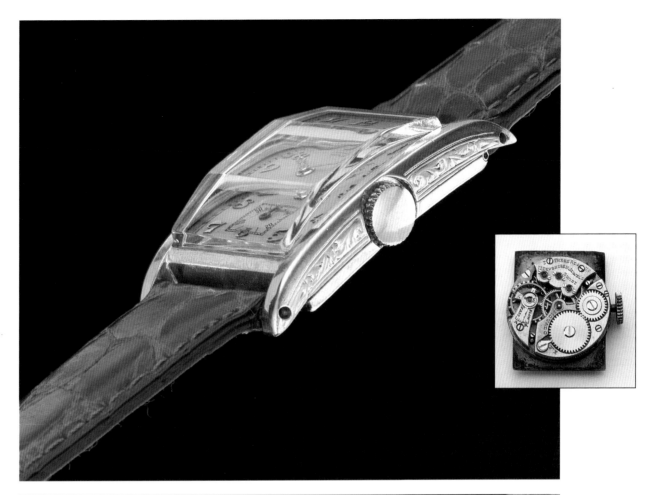

70 a, b, c
Rectangular, white gold cased gentleman's wrist watch with prism-cut glass, made in 1925. The case has engraved decoration on the band and a sunken winding crown. Silvered dial with small seconds, signed "Chronometer Movado", Reference No. 0790211. Nickel-plated, 8¾''', Calibre 780 movement with lever escapement, 17 jewels, 3 adjustments, 3 gold chatons with screws, signed "Movado Fy".

71 a, b
Lady's tonneau-shaped wrist
watch in diamond-set platinum case, made in 1925.
Silvered dial without
seconds, signed "Movado",
Reference No. 20 052. Form
movement, Calibre 11,
signed "Movado Fy",
15 jewels, 4 adjustments,
3 gold chatons with screws.

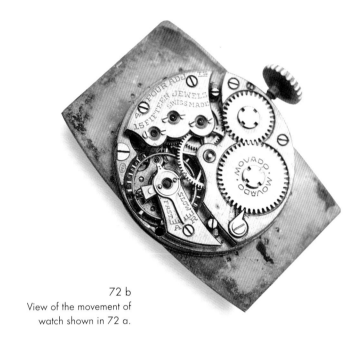

72 b
View of the movement of
watch shown in 72 a.

72 a
Tonneau-shaped 18 ct gold
wrist watch, made in 1930.
Silvered dial with small
seconds, signed "Movado",
Reference No. 74503591.
Round nickel-plated 8¾''',
Calibre 730 movement,
signed "Movado", 15 jewels,
4 adjustments, 3 gold
chatons with screws.

73
Rectangular 18 ct gold wrist
watch, made in 1930.
Silvered dial with small
seconds, signed "Movado",
Reference No. 746181 65.
The case back is engraved
with the inscription "Recuerdo
de Empleados De Sun Life en
Guatemala Agusto 1930".
Nickel-plated, round, 8¾''',
Calibre 780 movement with
lever escapement, 17 jewels,
4 gold chatons with screws.

74
Extravagantly shaped 14 ct
gold lady's wrist watch with
gold link bracelet, made in
1920. Silvered dial with
small seconds, signed
"Chronomètre Movado",
Reference No. 793065 452.
Nickel-plated, 8¾''', Calibre
780 movement with lever
escapement, 15 jewels,
4 adjustments.

The 1920's: "Valentino" and "Ermeto"

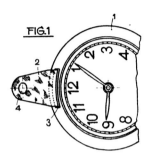

FIG.1

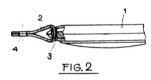

FIG. 2

75
Patent No. 135 524 for the attachment of the pocket watch version of the "Valentino".

76
Patent No. 122 390, dated 16th September 1927, for the pocket watch version of the "Valentino".

The fêted American star of silent films, Rudolf Valentino, died in 1926, aged only 31. His funeral took place with the sympathy of the whole Western world. For Valentino, with his Italian origins, represented the embodiment of elegance, the idol of the twenties. One year before his tragic demise, Movado had honoured him with two flat watches, a pocket and a wrist version, called "Valentino".

The pocket watch version of the Valentino generally had the case covered with snake skin, according to a patent applied for on 3rd December 1926 but only ratified on 16th September 1927. A flat, well-protected winding crown was integrated into the case at the 6 position, with the small seconds therefore at 12. Opposite the winding crown was a small round opening for a watch chain swivel. With a further patent, dated 2nd December 1929, the usual leather loop was fitted (mentioned as an alternative in the first patent) to which the watch could be buttoned, or equipped with a leather strap. Therefore this watch did not have a protruding ring, or pendant, and was thus soft, round and without corners. The Valentino was the original pattern for a particular form of pendant watch that was still produced in the 1940's as a so-called "Nurse's watch". Presumably this type of watch was so named by Movado and others as it was especially popular with nurses. There was also a version with a glass back.

The wrist watch version of the Valentino was patented three years later on lst September 1930. It had two opposed loops to take a woven leather strap. The case could also be leather covered, as with the pocket watch.

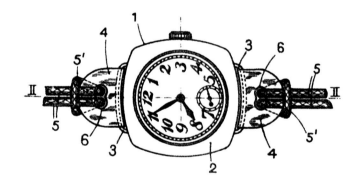

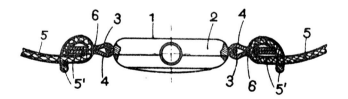

77
Patent No. 140 801 of 1st September 1930 for the wrist watch version of the "Valentino".

78
A post-1930 wrist watch version of the "Valentino" model. Silver case covered with green snakeskin. Silvered dial with small seconds, signed "Chronomètre Movado", Reference No. 21836 160152. The 10¼‴, Calibre 150 MN movement has a lever escapement, 15 jewels, 4 adjustments and 3 gold chatons with screws.

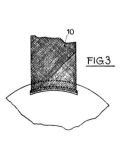

FIG.3

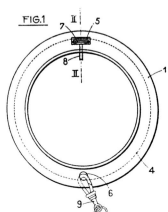

FIG.1

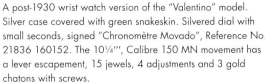

FIG.2

79
A pocket watch version of the "Valentino" model, made in 1926. Silver case covered with green snakeskin. Small seconds at the 12 and winding crown at the 6, Reference No. 27350 632903. Nickel-plated 16¼''', Calibre 620 movement with lever escapement, 15 jewels, 4 adjustments and 3 gold chatons with screws.

80 a, b
Elegant cushion-shaped gold Movado pocket watch entirely covered with enamel painting, made in 1925. This watch is a good example of an Art Deco watch case, Reference No. 251928. Nickel-plated, 17''', Calibre 250 movement with lever escapement, 19 jewels, 8 adjustments (precision adjusting) and 3 gold chatons with screws. Watches of this type were created for the exhibition "Exposition des Arts Décoratifs" in Paris in 1925.

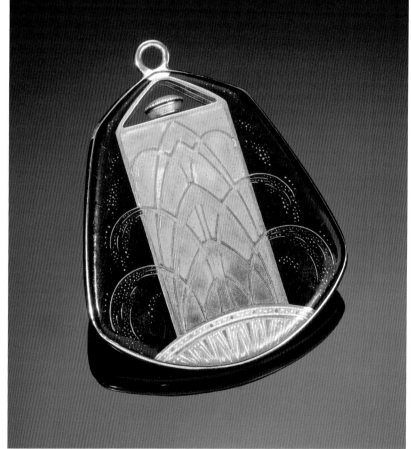

81
Pendant watch (so-called Nurse's Watch) based on the Valentino design, made in 1940. Steel case with sunken winding crown, multi-coloured dial, signed "Chronomètre Movado", Reference No. 11 709 0193588. Nickel-plated 10¼''', Calibre 150 MN movement with lever escapement, 15 jewels, 3 gold chatons with screws. The back of the watch is covered with leather.

82
Pendant watch (so-called Nurse's Watch) based on the Valentino design, made in 1940. Steel case with sunken winding crown, dial with centre seconds, signed "Chronomètre Movado". 10¼''', Calibre 150 MN movement with lever escapement, 15 jewels and 4 adjustments. The case has a glazed back.

83
Elegant gold Movado pocket watch with mother-of-pearl dial and gold numerals, small seconds, signed "Movado", made in 1920. Nickel-plated 16¼''', Calibre 620 movement with lever escapement and 17 jewels. (Private collection).

84 a, b
Movado pocket watch with chronograph and 30-minute counter, made in 1925. The silvered dial has small seconds and is signed "Movado", Reference No. 206229. The 17''' movement is presumably Valjoux Calibre 54, with slight modifications, signed "Movado".

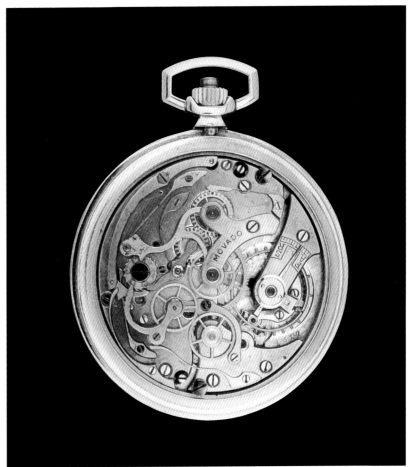

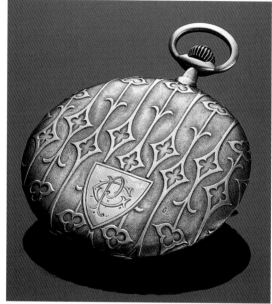

85 a, b, c

Double cover hunter-cased silver Movado pocket watch, the case decorated with niello, made in 1920. Silvered dial, with small seconds, signed "Movado Sûreté", Reference No. 7506, Serial No. 640405. Nickel-plated, 16¼''', Calibre 640 movement with lever escapement, 15 jewels, 3 gold chatons with screws, signed as the dial. Precision regulator index according to Patent No. 37 776. The case construction with accessibility from the front is according to Patent No. 51 670.

86 a. b, c
Finely decorated 18 ct gold
hunter-cased Movado pocket
watch, made in 1925. The
back cover of the 36.8 mm
diameter case has a champlevé technique enamelled
monogram. The silvered dial
with small seconds is signed
"Movado", Reference No.
695 050. Nickel-plated,
16¼''', Calibre 680 move-
ment with lever escapement,
15 jewels, 4 adjustments, 3
gold chatons with screws and
a simple regulator index.

87
Two delicately decorated Art Deco Movado gold pocket watches, made in 1925. The upper watch with champlevé technique enamel decoration, the lower octagonal watch with off-centre dial. Nickel-plated, 10¼''', Calibre 150 movements.

88 a, b
Elegant, flat gold pocket watch with enamel decoration, made in 1925. Silvered dial with gold numerals, small seconds, signed "Movado", Reference No. 42508-251345. Nickel-plated, 17''', Calibre 250 movement with lever escapement, 19 jewels, 8 adjustments (precision adjusting) 3 gold chatons with screws, precision fine regulator index according to a patented Movado design.

89 a, b
Delicately decorated Art Deco 18 ct gold Movado pocket watch, made in 1925. Unusually shaped bow, the winding crown and pendant form the head of an Egyptian sphinx. Silvered dial with small seconds, signed "Movado", Reference No. 7119-0634088. Nickel-plated, 16¼''', Calibre 620 movement with lever escapement, 15 jewels, 3 gold chatons with screws, patented balance spring stud and precision regulator index, signed "Movado", "filets" decoration.

90
Art Deco 18 ct octagonal gold pocket watch, made in 1925. The winding crown and pendant in the form of an Egyptian sphinx head. Gilded dial signed "Movado", with small seconds. Nickel-plated movement with lever escapement, 15 jewels, 4 adjustments, 3 gold chatons with screws and patented precision regulator index. The movement has "filets" decoration.

91
Cushion-shaped Art Deco pocket watch with 18 ct white gold and onyx case, made in 1925. Silvered dial without seconds, signed "Chronomètre Movado". Nickel-plated, 8¾''', Calibre 730 movement with lever escapement and 17 jewels.

Ermeto

One year after the Valentino, Movado began to market a watch enclosed by a rectangular case and called the "Ermeto". This daring watch design deviated even further from the typical pocket watch concept than the Valentino. With respect to unconventionality the Ermeto was, following the Polyplan, a second great success. It was advertised, with delightful modesty, as opening the third era of watchmaking history, following the era of the pocket watch and the age of the wrist watch.

The name "Ermeto" comes from the Greek and originally meant sealed (against the ingress of air and water). This group of watches, also called "closed pocket", "cased" or "golf watches", are described as hermetic watches, although they are in no way air or water tight. The designation "Ermeto" for the Movado model is only intended to suggest protection of the watch against external influences. This is certainly true and the watch is vastly better protected than it is with a normal pocket watch case (see also: "Hermetic" Watches, a review by Bernard U. Bowman Jr., NAWCC Bulletin No. 290, June 1994).

The Ermeto is a watch contained by a rectangular case. The movement capsule, comprising the movement, dial, winding and hand-setting mechanism, forms an autonomous unit which is enclosed by a two-part metal case. The two portions may be drawn apart, like curtains, to reveal the dial and allow the time to be read. The fact that both hands are required to do this was cause for criticism of the Ermeto. When the two halves of the outer case are closed, the movement capsule is given a certain measure of protection against shocks and dust, as well as air pressure and temperature changes. The Ermeto was developed by Huguenin Frères, Le Locle, and patented in Germany on 12th October 1926, DRP No. 443 555. Huguenin offered the idea for production to numerous watch manufacturers but only Movado took it up. Hugenin had foreseen the concept as a pendant watch but, after some modifications, Movado launched the Ermeto on the market with a lavish advertising campaign in 1926 –

92
DRP No. 443 555, dated 12th October 1926, issued to the Swiss company Huguenin Frères in Le Locle for the concept of the "Ermeto".

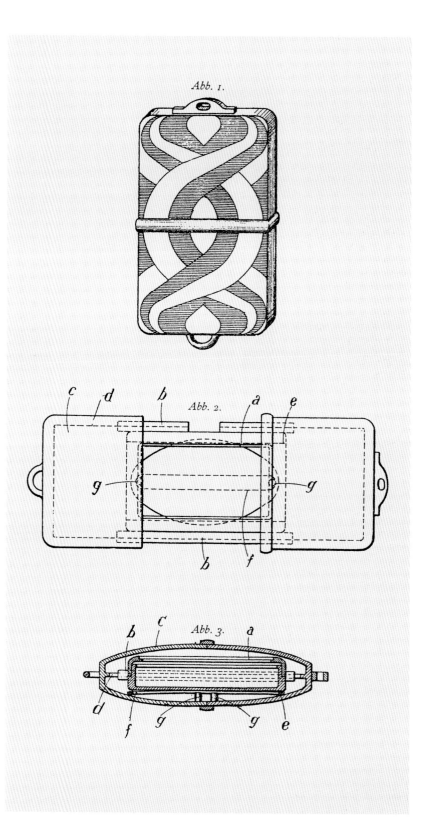

93
The Ermeto in 3 different sizes and versions; at the top the "Ermeto Pullman" as a travelling clock with 8 day movement. Centre, the "Ermetophon" with alarm, shown full size, made in 1960. Below, the "Ermeto Calendine" with date. The movement is shown in Illus. 114. (The "Pullman" is in private ownership).

Movado's first big campaign to promote a new product.

The magnitude of the market success of the Ermeto, or at least its potential success, may be judged by the fact that the marketing rights were, in themselves, treated as goods. A businessman called César de Trey, who formerly had a dentures business in England, returned to Switzerland in 1927, with the intention of going into the watch business. He acquired the sales rights for the Ermeto from Movado for the markets in France, England, Spain and Italy for SFr. 250,000.–. In 1928 he formed a company in Lausanne, which was called Hermética S.A., whose sole purpose was to market the Ermeto.

Originally the Ermeto was manually wound by means of a winding crown positioned at the 12. By about 1927 the idea was modified to include an automatic winding system that is simple, clever and reliable. As the two halves of the outer metal case are moved back and forth, two racks, one fitted in each half, engage a pinion on the winding stem and cause the latter to turn. Since the mechanism is completely enclosed, the winding crown turns, by itself, as if by magic (Illus. 107). A single opening operation provides sufficient winding of the mainspring for 4 hours running. Thus if the case is opened six times the watch will run for 24 hours. The German patent for the rack-winding system was issued to Movado on 29th February 1928. The equivalent Swiss patent, No. 127 820, dates from 1st October 1928. When the alarm version of the Ermeto (Ermetophon) was introduced in 1955, the upper edge of the watch was changed into the form of an arc with two winding crowns, one for the going train and one for the alarm. Because of this, the construction of the automatic winding system had to be changed. It functions by means of two side levers, operated as before by the two halves of the outer case (Illus. 115).

There were four different sizes of Ermeto. The largest was called "Pullman", from the American express train of the same name, and was intended as a travelling clock. It had either a 19''' or 21''' eight-day movement, sometimes with alarm. The next largest size, the "Master", was robustly laid out and was equipped with a 12''' manually wound movement. Then came the "Normal" size, usually with the calibre 150 MN movement, either automatically or manually wound. The smallest of all was called the Ermeto "Baby" and was principally intended for ladies' handbags. The movement was a 5¼''' size, once again either automatically or manually wound. Apart from these 4 sizes,

94 a–f
Designs for various Ermeto
enamel cases from approximately 1935.

77

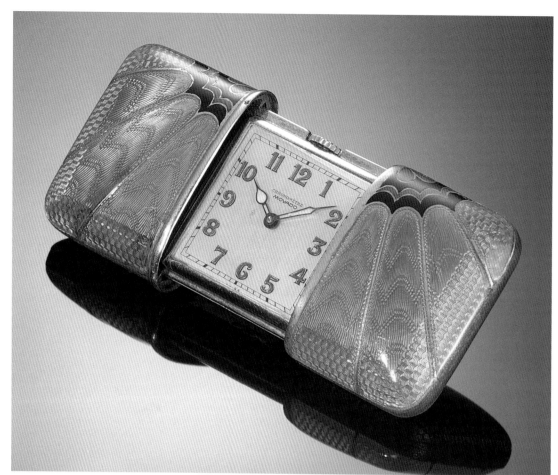

95
Blue enamelled and decorated Ermeto for the Spanish market, made in 1930. The silvered dial without seconds is signed "Movado Suiza Carlos u. Perret Rosario". Calibre 150 MN movement with lever escapement. ▽

96 a, b
Extremely precious multi-coloured Ermeto with cloisonné enamel and engine turning, made in 1925. The silvered dial without seconds, signed "Chronomètre Movado". Calibre 150 MN movement with lever escapement.

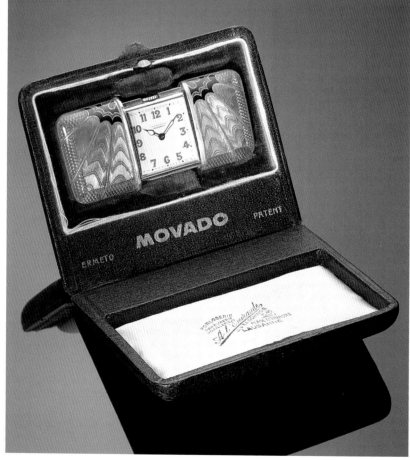

97
Three different executions of the "Ermeto". Above, a "Chronomètre Movado", made in 1930. The case is a rare gilded silver version with niello decoration. Calibre 150 MN movement. Centre, a "Calendermeto" signed "Cartier", made in 1950, with a reptile skin-covered steel case and Calibre 155 movement. Below, an "Ermeto" with a touch dial for blind users, made in 1940, with a leather-covered steel case and Calibre 150 MN movement.

98
Advertising material for the
Ermeto from early 1930.

99
English advertising leaflet for the Ermeto from early 1930.

See Illus. 260

100
Double page advertising leaflet for the Ermeto from early 1930.

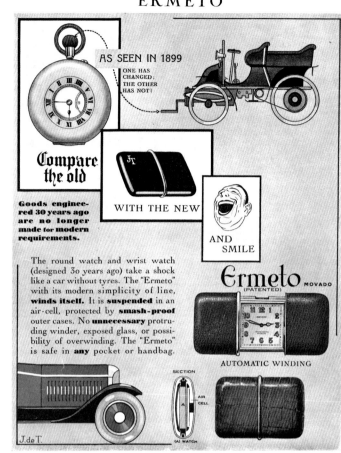

ERMETO

AS SEEN IN 1899

ONE HAS CHANGED: THE OTHER HAS NOT!

Compare the old

WITH THE NEW

AND SMILE

Goods engineered 30 years ago are no longer made for modern requirements.

The round watch and wrist watch (designed 3o years ago) take a shock like a car without tyres. The "Ermeto" with its modern simplicity of line, **winds itself.** It is **suspended** in an air-cell, protected by **smash-proof** outer cases. No **unnecessary** protruding winder, exposed glass, or possibility of overwinding. The "Ermeto" is safe in **any** pocket or handbag.

Ermeto MOVADO
(PATENTED)

AUTOMATIC WINDING

SECTION
AIR CELL
(A) WATCH

J.de T.

NON — FRAGILE

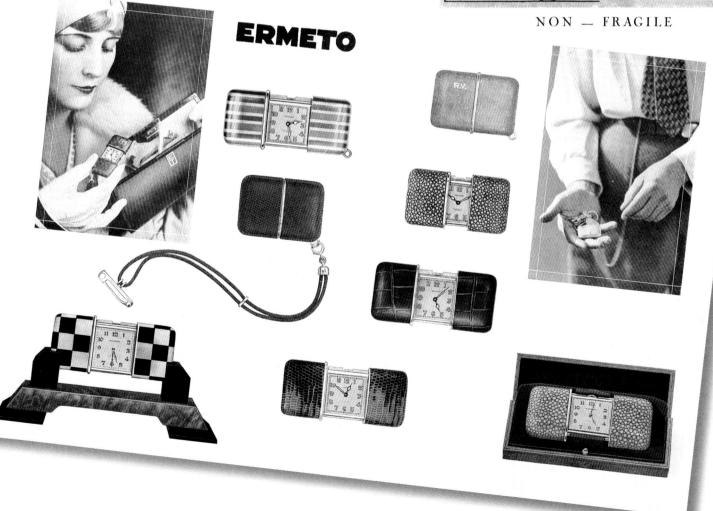

ERMETO

101 a, b
An "Ermeto Bag", to be integrated in the strap of a lady's handbag, made for Tiffany in 1935. The bag and the watch case are both covered with brocade. The dial is signed "Tiffany & Co". The movement is the small form Calibre 11.

102
Two "Ermetos" in coloured enamel cases, made in 1945. The upper watch, with black dial and gold numerals, is signed "Movado Cartier". The lower watch is unsigned. Both have Calibre 150 MN movements.

103
Two watches made by Movado especially for the 1958 World Exhibition in Brussels. Left, a square 18 ct gold wrist watch with calendar (day of the week and date in apertures), signed "Bruxelles 58", Reference No. 8477 213 899. Round 11¾''', Calibre 118 movement with lever escapement, 17 jewels and automatic winding. Right, a "Calendermeto" with the case covered in crocodile leather, likewise signed "Bruxelles 58", Reference No. 1262 020 M, silvered dial with date chapter ring, day of the week, month and phases of the moon shown in apertures. 10½''', Calibre 155 movement with lever escapement and 17 jewels.

104
"Ermeto Luxe" in an 18 ct
gold case. The dismantling of
the two movable case cap-
sules is illustrated. After turning the two screws in question about 120° counter
clockwise, the stop is freed
and the capsule may be
removed.

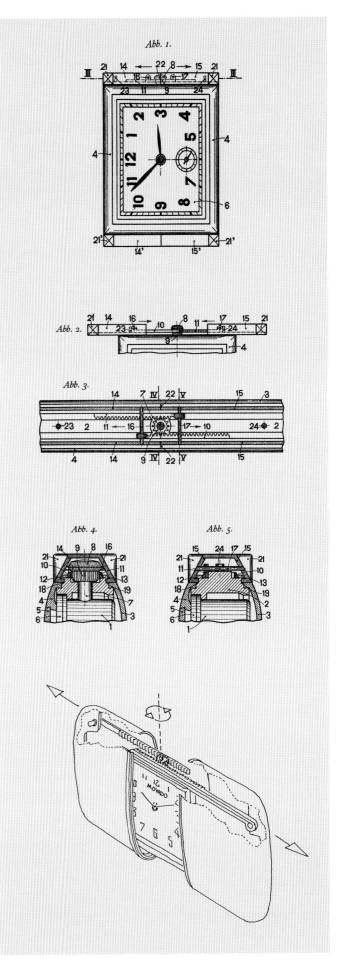

106
Movado Deutsches Reichs-
patent of 1928 concerning
an automatic rack-winding
mechanism similar to that of
the Ermeto.

107
Automatic winding of the
normal Ermeto models.

numerous variations were made over the years (see general view of the model). The Ermeto was made with alarm, calendar, centre seconds, and even as a chronometer. This may be defined as having a certificate, following rating tests, that was equivalent to the standards of the official Swiss Observatory test (B.O.). Some particularly finely adjusted Ermetos were said to be entered for Observatory Chronometer Trials, following particularly careful rating.

The Ermeto could be carried loose in the pocket or attached to a chain, like a pocket watch. It was equally suitable for the elegant lady's handbag: in 1931 a version named "Ermeto Bag" was even developed to be built into the strap of a lady's handbag. And a more robust version was conceived for a hiker's rucksack. Thus Ermeto soon became known as the only unisex watch, suitable for either men or women. This misconception corresponded to the earlier observed preference of Movado for medium-sized wrist watches, which could serve as ladies' or gentlemen's models. Eugène Jaquet and Alfred Chapuis celebrated the success and ingenious concept of the Ermeto by devoting a chapter to it in their "Histoire et technique de la montre suisse", published in 1945. The Ermeto was manufactured until 1985, including those produced by Zenith from 1972 to 1983, when that firm owned the rights to the Movado name.

As conventional as the basic mechanism of the Ermeto was, it presented new and various possibilities to adapt its shell-like outer case to suit almost any purpose. This adaptation could be achieved by employing different case materials and decoration. These possibilities were exploited as widely as possible. Cases of gold or silver could either remain smooth, or be engraved or engine turned in various ways. Cases that were not made from precious metals could be gilded (likewise those from silver), silvered, or covered with expensive leathers such as crocodile, lizard, ostrich, shark and snakeskin. The case could also be coloured with Chinese lacquer, covered with enamel, or even set with precious stones. Thus the Ermeto became a piece of jewellery, a work of art in the Art Déco style, which prevailed both in Europe and the USA at the time of its creation.

The majority of the Ermeto cases were made by casemakers in Geneva and a range of decorative designs has survived (Illus. 94). A brochure from the sales company, Hermetica S.A., from about 1930 lists 30 various available executions, made in series (refer to the list). The non-precious metal versions are not included, possibly because they had not yet been made at that time. The price list shows that the "automatically" wound version was about a third more expensive than the manually wound one. It also shows that the price range of the highly-developed versions could be as much as 6.5 times the price of a simple one and this without including the luxury models with precious stones or platinum cases. It is no wonder such firms as Cartier or Tiffany included the Ermeto in their line of products, creating beautiful and valuable cases for it.

The Movado Ermeto, which won the Grand Prix at the World Exhibition in Barcelona in 1929, must be considered the absolute prototype of all the "hermetic" and purse watches, taking into account its success and level of fame. Many companies produced imitations of the Ermeto; more than 40 various models are known but were never able to repeat Movado's success. Most manufacturers, however, sought to avoid the one disadvantage of the Ermeto; namely that it required both hands to operate it.

The Ermeto, a view of the various models

Model	Year introduced	Calibre	Remarks
Ermeto	1926	150 MN	
Ermeto	approx. 1935	157	
Ermetolux	approx. 1930	105	
Ermeto Pullman	approx. 1930	310 M	8-day movement
Ermeto Pullman	approx. 1935	711	Lemania, as above, with alarm
Ermeto Pullman	approx. 1945	895	AS calibre, 8-day with alarm
Ermeto Bag	1931		Incorporated into handbag
Ermeto Calendine	approx. 1945	578	Form movement with calendar
Calendermeto	1948	155	With calendar
Baby Ermeto	approx. 1940	575	Form movement
Ermetophon	approx. 1955	900	AS calibre 1475, alarm
Ermetophon	approx. 1955	901	AS calibre 1475, alarm
Ermeto Luxe (Luxury)			18-ct. gold, engine-turned gold case.
Ermeto Standard			Manual winding

108
Ermeto advertising leaflet
from about 1932.

109
Another Ermeto advertising
leaflet from about 1932.

ERMETO

HERMETICA S.A.
LAUSANNE (SUISSE)

LE
TROISIÈME AGE
DE LA MONTRE

ermeto

ERMETO
le troisième âge
de la montre

ERMETO
la montre d'aujourd'hui
et de demain

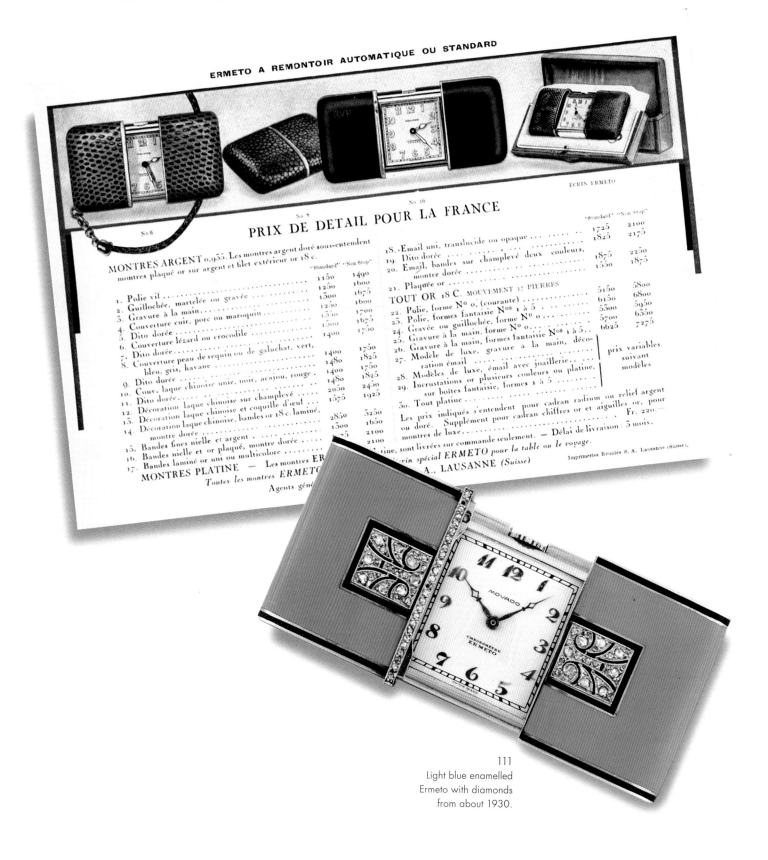

110
French price list for the
Ermeto from about 1932.

111
Light blue enamelled
Ermeto with diamonds
from about 1930.

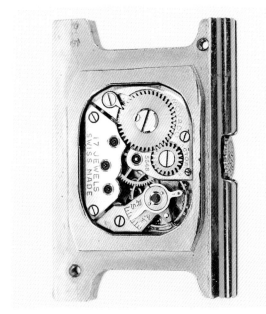

In the mid 1920's, in 1924, it appears that a first American Movado branch office was set up in New York by Gaston Ditesheim with the collaboration of Edwin Hellinger. The latter was of Hungarian descent and an experienced manager who had been trained in Geneva. With his good taste and marketing acumen, Hellinger played a decisive role with regard to the creation of new models, and it was he who gave the head office in La Chaux-de-Fonds the tip of dedicating a watch to Rudolf Valentino. The existence of the New York office is first evidenced in writing in 1934 with an advertisement in the June/July issue of "Horology", in which watchmakers are welcomed to the "Movado Watch Agency Inc." at its new business premises at the Rockefeller Centre, 610 Fifth Avenue.

The close of the 1920's also marked the entry of a new generation at the head office in Switzerland. The Ditesheim sons began to replace the founding fathers, now in their sixties, on the board of directors. First was George, son of Isaac (1860 to 1928), who had graduated from the School of Horology in La Chaux-de-Fonds with a Diploma in Horology and had already joined the company in 1912. Then Pierre, one of Achille's six children, who took the place of his 67-year-old father. The remaining brother was the 62-year-old Isidore, who still held the reins of the board of directors, and who once again received the honour of serving on the jury for the 1930 World Exhibition at Liège.

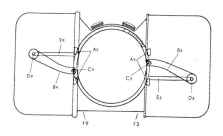

It has already been mentioned that the Ditesheim brothers first began to take part in the chronometer trials at Neuchâtel Observatory in 1899. They entered with a pocket watch movement signed "Apogée", which was presumably based on an ébauche purchased from another manufacturer (Illus. 11, page 18 f.). This illustration, in a 1948 Movado brochure, together with a copy of the rate certificate (Bulletin de marche), are the first and only clues to the Ditesheim brothers' entry into chronometry. From the certificate it may be seen that this chronometer by L., A. & I. Ditesheim with movement No. 1904 had a lever escapement and a balance spring with double Phillips terminal curves. We also learn from this source that the Ditesheim company entered 26 watches in the 1899 trials, 22 of which were awarded certificates, six of these receiving first prizes.

It was also in 1899 that the official Swiss Watch Testing Office (B.O.) in St. Imier registered 30 rating certificates for Ditesheim pocket watches. This office (one of the then four, later seven, official Swiss watch testing offices) subjected watches to a rigidly laid down testing procedure and issued "Chronometer" certificates to those watches which ran within the prescribed rates. Of the 30 Ditesheim pocket watches which received certificates, four took places between 8th and 11th position, out of a field of no fewer than 582.

The tests and rate certificates of the two Swiss observatories, Neuchâtel and Geneva, differ noticeably from those of the B.O. The observatory tests are longer and the certificates are much more difficult to attain. In addition, the chronometers that had achieved a rate certificate from an observatory, with the title Observatory Certificate ("Chronomètre d'Observatoire"), rather than a B.O. certificate, were entitled to participate in an annual competition with the chronometers from other manufacturers and could receive various prizes. This was thus a true contest between Swiss precision watch manufacturers, open to all-comers, as the competition results were published. With the B.O., on the other hand, there was

116 a, b
Movado pocket watch for adjusters with only a small seconds dial. Steel case, dial signed "Coincidence". Simple lever escapement movement with an externally operated balance stop, 15 jewels, simple regulator index, signed "Movado".

no competitive element; the aim was simply to achieve a sales promoting rating certificate. The chronometers tested by the B.O. were always intended to be sold. In the case of "Chronomètres d'Observatoire", however, some were intended for sale, their value being enhanced by a "Bulletin d'Observatoire", enabling them to command a higher price, while others belonged to a special series reserved for competitions and would only be sold in exceptional circumstances. In Switzerland, Movado only took part in the Observatory competitions in Neuchâtel, not in Geneva, on a regular basis between 1912 and 1939. In the period 1931 – 1935, the worldwide depression years, however, no watches were entered.

Up until 1945 there were five test categories, with differing test conditions, at Neuchâtel:
1. Marine chronometers, size unrestricted
2. Deck watches larger than 60 mm movement diameter
3. Deck watches up to 60 mm movement diameter
4. Pocket chronometers larger than 45 mm movement diameter
5. Pocket chronometers with a movement diameter between 38 – 45 mm

Beginning in 1945, wrist chronometers were added, which will be discussed at a later point. Up until then, Movado took part in the deck watch and pocket chronometer categories, with tests lasting 55 and 46 days respectively. Only once, in 1922, did Movado enter in the marine chronometer category, when three especially constructed deck watches were submitted.

On 6th July 1946, the Director of Neuchâtel Observatory notified Movado that the company had qualified 21 times for the series prize with the 6 best chronometers in the period 1912 – 1941. This means that in only 6 years of the 27-year cycle (up to 1939). Movado had not won this prize. Moreover, this includes the 4 years, 1931 – 35, when Movado did not enter the trials. It may be ascertained from other sources that, apart from this, Movado won a total of 165 first prizes and a row of second and third prizes in the period up to 1939. Movado's regular participation in the Neuchâtel trials ended, as we have said, in 1939. The company only took part once again, in 1948. This was in an international competition, held in addition to the normal trials, to celebrate the centenary of the foundation of the Republic of Neuchâtel. Movado won a series prize, two firsts, a second and a third prize. Previously, in 1923, another peak had been reached in a special competition to celebrate the centenary of the death of

Abraham Louis Breguet, the famous watchmaker from Neuchâtel. Movado won a series prize and two first prizes. The "Livre d'or de l'horlogerie", published in Geneva and Neuchâtel in 1926 to mark the 50th anniversary of the Journal Suisse d'Horlogerie, states the following concerning Movado's chronometry activities:

"(Fabriques Movado) has been the most frequently mentioned company throughout the years, at home and abroad, in connection with observatory successes. They have won the series prize at Neuchâtel Observatory every year since 1912. Last February they won the overall lst prize in the deck watch category."

However, Movado sought not only to measure their precision performance against their competitors in Switzerland, but also those at the English National Physical Laboratory (NPL) at Teddington, which had taken over the duties of rating and competitive trials from Kew Observatory in 1912. At Teddington no chronometers were awarded prizes, but rating certificates were issued for two classes, A and B. As in the Swiss observatories the watches were also awarded a number of points corresponding to the rating achieved, a decisive factor for their placing in the competition. A "Class A Certificate" from either Kew or Teddington represented considerable international prestige value when it was also connected with a classification in the upper placings. Movado took part in the Teddington competitions from 1918 on, and in 1923 and 1930 gained lst place with movements of their large calibre 355. In the 1930's they belonged to the group of seven Swiss watch manufacturers – along with Omega, Zenith, Ulysse Nardin, Patek Philippe, Longines and Paul Ditisheim – which had completely dominated these competitions since 1920. For example, of the 25 best deck watches for 1936, in which the Movado Calibre 355, No. 360515, attained 7th place, there was no longer a single English watch; the winners were all of Swiss origin. All the Movado watches entered at Teddington had previously competed at Neuchâtel and only the best ones were sent to England.

For their deck watches and pocket chronometers Movado employed three different calibres, or series of movements.

1. The first is a calibre with the usual Movado characteristics, the cock for the train with three rounded projections and a movement diameter of 43 mm (19‴). The watches of this calibre are numbered from 360 001 – 360 045 and were used between 1912 and 1922. They should not be

117

Movement of a Movado Observatory pocket chronometer from the 1st Chronometer Series fitted into a wooden ring for use as a display watch, made in 1920. Nickel-plated, 19‴, movement, No. 360 032 with "filets" decoration, signed "Movado Fy", lever escapement, 21 jewels, 4 gold chatons with screws, Guillaume balance, flat balance spring with double Phillips terminal curves and patented precision regulator index. This watch successfully took part in the pocket watch category of the Chronometer Trials at Neuchâtel Observatory in 1920.

confused with the normal pocket watch calibre, of similar size, also from this era, whose six-digit movement numbers also commence with 360. In addition to the already high quality execution of the normal calibre with its precision regulator, train jewels with chatons, end stones for the escape wheel and balance, the observatory watches had Guillaume balances and balance springs with double Phillips terminal curves. Most of the observatory series are nickel plated and decorated with côtes de Genève stripes, although this decoration is a later addition. For the trials they were plain, without this finish, and had flat-stoned brass plates and bridges with the corners simply broken.

Taking into account their small size (compared with those of the competitors) the total of 45 watches which comprise this series amassed an amazing number of prizes up to 1922. Particularly bearing in mind the fact that they are merely up-graded series movements, these pieces achieved remarkably good rates. One movement from this series, No. 360 032 (Illus. 117) was subsequently fitted into a wooden display case. The engraving on the broad spacing ring, "ler Prix", cannot be correct; although this watch was awarded a certificate in July 1920, it did not win a first prize.

2. The second calibre, like the first, has a typical Movado form but is 50 mm diameter (22‴) and at first sight looks like an enlarged version of the former. It is numbered from 360 434 – 360 478 and may be traced to the period 1919 – 1927. These movements are equipped as follows: a large diameter (22 mm) Guillaume balance, balance spring with double Phillips terminal curves, a new pallet bridge shape with endstone, a total of 21 jewels, and a simple index with fine adjustment for the curb pins. The jewels are mostly set directly in the plates, only occasionally in a chaton, as for example the escape wheel arbor of No. 360 463 wheel arbor (Illus. 119).

The rates of these movements in trials were markedly superior to those of the first series. It was watches with this calibre that achieved the success in the special Breguet competition of 1923. There were apparently two watches of this calibre, but with different numbering, equipped with spring detent escapements, and it is highly probable that the only traceable Movado tourbillon, No. 360 433, was one of this series. These three watches are now missing.

3. The third calibre has a completely different form. It has a distinctly English appearance with a three-quarter plate, and a diameter of 65 mm (28‴). It has been traced to the period 1927 – 1939, which is right up to the end of Movado's regular participation at Neuchâtel. The very plain appearance of this calibre points to the fact that it was developed entirely with stability and highest precision in mind. The very large 26 mm diameter balance is pivoted beneath a large and stable balance cock. Apparently a bimetallic compensated balance was used at first, but this was later replaced by a Guillaume balance, as in the first series. Once again the balance spring has double Phillips terminal curves. The barrel is placed under a bridge that may be removed separately. Likewise the escapement is fitted on a plate of its own which makes it relatively simple to convert a movement from the normal lever escapement to spring detent, as shown by some examples of this calibre. The curb pins have a fine adjustment and the balance staff a diamond endstone. Most

118

The rating certificate of pocket chronometer No. 360 032 issued by Neuchâtel Observatory (see Illus. 117). The watch is now mounted in a display case.

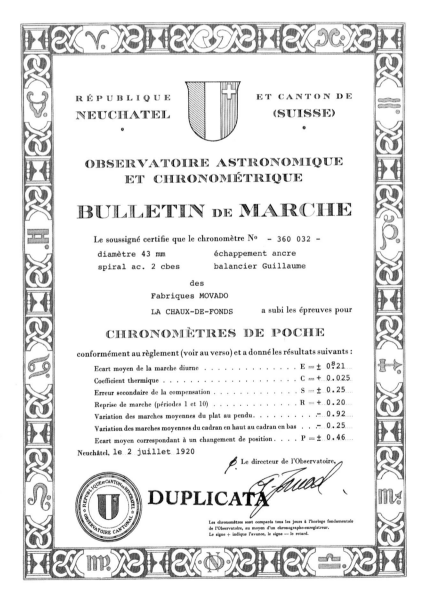

RÉPUBLIQUE NEUCHATEL · ET CANTON DE (SUISSE) ·

OBSERVATOIRE ASTRONOMIQUE ET CHRONOMÉTRIQUE

BULLETIN DE MARCHE

Le soussigné certifie que le chronomètre N° – 360 032 –
diamètre 43 mm — échappement ancre
spiral ac. 2 cbes — balancier Guillaume
des
Fabriques MOVADO
LA CHAUX-DE-FONDS — a subi les épreuves pour

CHRONOMÈTRES DE POCHE

conformément au règlement (voir au verso) et a donné les résultats suivants :

Ecart moyen de la marche diurne . . . . E = ± 0^s21
Coefficient thermique . . . . C = + 0.025
Erreur secondaire de la compensation . . . . S = ± 0.25
Reprise de marche (périodes 1 et 10) . . . . R = + 0.20
Variation des marches moyennes du plat au pendu. . . . . – 0.92
Variation des marches moyennes du cadran en haut au cadran en bas . . – 0.25
Ecart moyen correspondant à un changement de position. . . . P = ± 0.46

Neuchâtel, le 2 juillet 1920

Le directeur de l'Observatoire,

DUPLICATA

Les chronomètres sont comparés tous les jours à l'horloge fondamentale de l'Observatoire, au moyen d'un chronographe-enregistreur. Le signe + indique l'avance, le signe – le retard.

des *Fabriques MOVADO LA CHAUX-DE-FONDS*

Épreuves pour chronomètres de poche

Année *1920*

No de Dépôt : P *I 51*

[Certificate data table — Observatoire Astronomique et Chronométrique de Neuchâtel, with columns for Périodes, Date, Marches diurnes, Écarts avec la moyenne de la période, Temp., for various positions: Position horizontale cadran en haut chronographe en marche; Position verticale pendant en haut chronographe en marche; Position verticale pendant en haut; Position verticale pendant à gauche; Position verticale pendant à droite; Position horizontale cadran en bas; Position horizontale cadran en haut; Position horizontale temp. 4°; Position horizontale; Position horizontale temp. 36°; Position verticale pendant en haut; and a RÉSUMÉ and COMPENSATION section.]

RÉSUMÉ

Périodes	Marches moyennes	Écarts moyens	Sommes des écarts	Temp.	Écarts de position	Positions
1'						horiz., cadran en haut
2'						vert., pendant en haut
1	-0.07	0.17	0.70	18.5		vert., pendant en haut
2	+0.30	0.46	1.60	18.3		vert., pendant à gauche
3	+1.07	0.22	0.90	18.5		vert., pendant à droite
4	+0.70	0.15	0.60	19.0		horiz., cadran en bas
5	+0.95	0.15	0.60	18.8		horiz., cadran en haut
6	+0.52	0.22	0.90	4.8		horiz., cadran en haut
7	+1.42	0.23	0.94	18.6		horiz., cadran en haut
8	+1.20	0.30	1.20	32.1		horiz., cadran en haut
9	-0.25	0.30	0.80	18.5		vert., pendant en haut
10	+0.75	0.07	0.30	18.8		vert., pendant en haut

COMPENSATION

Périodes	t	m	C = (m₃₆ - m₄)/(t₃₆ - t₄)	½[(m₃₆ - m₂₀) S/2 + (m₄ - m₂₀)]
6		m₄ =	m₃₆ - m₄ =	
½(5+7+9)		m₂₀ =	t₃₆ - t₄ =	m₃₆ - m₂₀ =
8		m₃₆ =		2 S =
			C =	S =

Ecart moyen de la marche diurne . . . . E = +0.21

Coefficient thermique . . . . C = +0.025

Erreur secondaire de la compensation . . . . S = +0.25

Reprise de marche (période 10 — période 1) . . . . R = +0.20

Variation des marches moyennes du plat au pendu ((pér. 1 + pér. 10)/2 — pér. 5) . . . . −0.92

Variation des marches moyennes du cadran en haut au cadran en bas (pér. 4 — pér. 5) . . . . −0.25

Ecart moyen correspondant à un changement de position . . . . P = +0.46

Variation des marches moyennes du plat au pendu, chronographe en marche (pér. 2' — pér. 1')

119

Two Observatory chronometer movements by Movado. Right, No. 360 011 from the 1st Observatory Series. Finely gilded 19''' movement with lever escapement, 4 gold chatons with screws, Guillaume balance, flat balance spring with double Phillips terminal curves and patented precision regulator index, signed "Movado Watch Fy". Left, No. 360 450 from the 2nd Observatory Series. The 22''', Calibre 360 movement has unfinished brass bridges and cocks and has no signature or other engraving. It has a lever escapement, Guillaume balance, flat balance spring with double Phillips terminal curves, simple regulator index with precision adjustment of the curb pins and a gold chaton with screws for the escape wheel arbor.

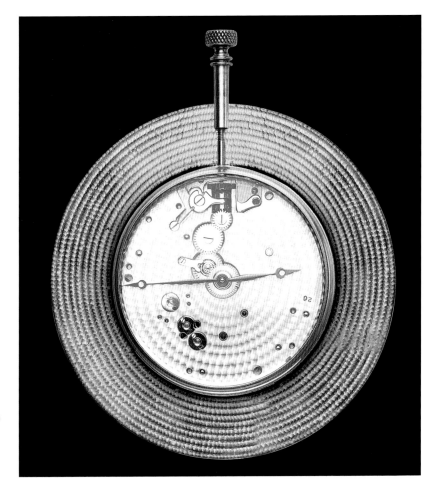

known examples of this calibre are gilded (presumably subsequently) and have been engraved "10 adjustments", pointing to a particularly thorough adjustment which is more comprehensive than the requirements for most European chronometer trials. The usual adjustments for chronometers were 7 or 8: in five positions (three vertical and two horizontal) and at two or three different temperatures (warm, ambient and cold). The two further adjustments of this calibre were probably the fourth vertical position (pendant down) and to the isochronism. Pocket watches with 10 adjustments, including these last two conditions, were tested at the former Deutschen Seewarte in Hamburg, the only European institute to do so. This calibre was thus also suitable for the tests in Hamburg.

This calibre has two series of numbers. The first is 355 001 – 355 009, whereby the first three digits signify calibre 355. The second series runs from 360 481 – 360 519, comprising 48 movements in all. 36 of them were tested in Neuchâtel, 35 of which were awarded 1st prizes in the pocket chronometer category. Out of the 35, a total of 18 subsequently also gained 1st prizes in the deck watch category. This gives clear proof of the undeniably high quality of this calibre.

Illustrations show a set of two watches of this calibre, No. 360 513 with lever escapement, and No. 360 507 with spring detent escapement. These watches are individually fitted into the usual simple hinged wooden boxes employed at Neuchâtel during the chronometer trials. These boxes may be locked with only the winding button projecting. Both boxes lie together in a padded wooden case. No. 360 513 was awarded a 1st prize at Neuchâtel in 1938. The version with spring detent escapement cannot be traced in the competition. It is remarkable that this is also the case with the other watches of this series fitted with spring detent escapements.

There is one known example of a deck watch of this calibre that has only been adjusted for temperature and not in positions, as is inferred by its inscription. Such insufficient adjustment would not allow this instrument to qualify at an observatory trial. It is also designated only as "Chronomètre de Bord" and not as "Chronomètre d'Observatoire". However, there may be some doubt as to the veracity of the inscription, relative to the adjustments, since this watch carries a combination of three letters on the balance cock which indicate that it was destined for export to the USA. Such inscriptions as "unadjusted, adj. to

120
Display watch fitted into a gilded brass ring. The watch is No. 360 460 from the 2nd Series and the under-dial finish is of the highest quality.

121
Movement side of a display watch with the highest quality finish: No. 360 460 from the 2nd Series. The movement is fitted into a wooden frame which may be hung on a wall.
(Privately owned).

temperature" indicating insufficient adjusting are frequently found on export watches, presumably for customs reasons.

A fourth type of pocket and deck watch movement may better be described as a movement layout rather than a calibre. It consists of a metal disc, which serves as a basic plate, on which is mounted a much smaller normal Movado calibre. The movement is positioned in such a way as to bring the fourth wheel axis exactly onto the centre of the plate and the dial. This construction makes a directly driven centre seconds hand possible, which is preferable to a small seconds hand for a precision watch. However, the centre wheel axis is then no longer at the centre, and since this must likewise be directly driven, it is usual for watches of this type to have an additional small, de-centralised chapter ring for the hours and minutes placed between the centre of the main dial and 12. The area below, between the centre of the dial and 6, carries an "up" and "down" indicator to show the reserve of winding available. This clever idea makes it possible to use smaller series movements, or even wrist watch movements, in a larger deck watch. Movado protected this idea by Swiss patent No. 104595 dated 22nd May 1923.

This type of watch has so far been seen in two sizes. The larger version employs a 43 mm diameter movement in a watch with a total diameter of 80 mm. The smaller one has a 10½‴ (23.6 mm) movement and an outside diameter of 46 mm, making it the size of a normal pocket watch, but with a wrist watch movement. The movement numbers so far recorded are between 340073 – 340109, thus belonging to a third series beyond the series commencing with 355 and 360. Possibly this series commencing with 340 was reserved for the combination movement layout. All watches of this type are placed in wooden boxes as navigational chronometers. They were not used as observatory chronometers since they are not identifiable in the competition records. Their construction does not lend itself to the required highest precision. They carry only the inscription "Chronometer" and have therefore presumably passed the chronometer tests at a B.O.

Who were the adjusters, with their difficult precision adjustment work, that enabled the watches to pass the difficult Observatory Trials and thus allow Movado to join the front ranks of the Swiss precision watch manufacturers so rapidly?

122
Two Movado deck watches in their original test cases. Above, No. 360 516 with "up" and "down" indicator. This watch is from the 3rd Observatory Series. Below, No. 360 463 from the 2nd Observatory Series.

123 a, b
Deck watch No. 360 516, as shown above, made in 1936. Gilded 28‴, Calibre 355 movement, lever escapement, signed "Movado Factories Swiss", 10 adjustments, 21 jewels, Guillaume balance, flat balance spring with double Phillips terminal curves, simple regulator index. Movement, 123 b (left).

Ernest Frey was the adjuster between 1912 – 1913. He worked in conjunction with Edmond Ditesheim, a cousin of the Movado founding generation, who was with Movado up to 1929 and was head of the adjusting department. Charles Hänggeli worked for Movado between 1914 – 1916 and the well-known Louis Augsburger was engaged by the firm in 1917. Augsburger was a freelance adjuster who also worked for Ulysse Nardin until 1932, when he went to work exclusively for Nardin. It was Augsburger who also adjusted all the Movado chronometers for the Breguet Competition in 1923. In 1936 the renowned adjuster Werner Albert Dubois joined the firm to continue the series of successes set by his predecessors. He also adjusted the first Movado wrist observatory chronometers at the start of the 1950's.

There was thus a total of 129 pocket chronometers and deck watches with which Movado competed very successfully in the chronometer trials and competitions at Neuchâtel and Teddington. Many of these watches were entered up to four times each. No. 360 436 was entered five times, taking a fifth at Teddington in 1921, at Neuchâtel a 2nd prize in 1919 and a 1st prize in 1921. Movado had turned to the production of precision watches at a very early stage with the development of the lever escapement for small ladies' watches and the numerous early "Chronomètres". This quest for high precision performance has been continuously and successfully pursued with extraordinary results.

Note: For this chapter I am indebted to Herr Herbert Neumüllers for his comprehensive and valuable information on the Movado observatory series. His essay was generously placed at my disposal before the foreseen publication in "Chronometrophilia". My grateful thanks are extended to him.

125
Two Movado Observatory chronometer movements from the 3rd Series, Calibre 355, made in 1938. Below, No. 360 513, gilded and finely finished, lever escapement. The movement, diameter 28''', signed "Movado Factories Swiss", 10 adjustments, 21 jewels, Guillaume balance, spring with double Phillips terminal curves, simple regulator index. This watch took part in the 1938 Chronometer Trials at Neuchâtel Observatory and was awarded a first prize in the pocket watch category. Above, No. 360 482, without finish, lever escapement, unsigned but to the same specifications as the watch below. This watch was awarded a third prize in the deck watch category at Neuchâtel Observatory in 1938.
(Privately owned).

124
Set of two Movado Observatory chronometers from the 3rd Series in a padded wooden box. Both watches, made in 1938, are in the hinged wooden cases used during the Chronometer Trials and have 28''', Calibre 355 movements. Left, No. 360 507 with spring detent escapement, right, No. 360 513 with lever escapement.
(Private collection, A. Simonin).

126
Detail of the Observatory chronometer No. 360 507 shown in Illus. 124. The movement, with spring detent escapement, is signed "Movado Factories Swiss", 21 jewels, 10 adjustments, Guillaume balance, flat balance spring with double Phillips terminal curves, simple regulator index. (Private collection, A. Simonin)

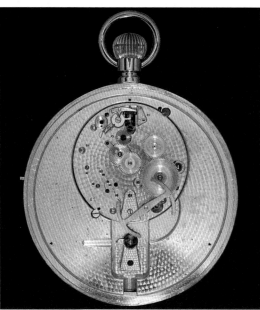

127 b

127 a

127 a, b, c
Large Movado deck watch (approx. 80 mm outside diameter) in a chromium-plated case, made in 1930. The silvered dial has a centre seconds hand, hours and minutes above the centre, an "up" and "down" indicator at 6 and is signed "Chro-nomètre Movado Chronome-ter". Reference is also made to Patent No. 104 595. Nickel-plated 19''', Calibre 350 movement with lever escapement, "filets" decora-tion. The movement is eccen-trically placed in a larger plate such that the fourth wheel arbor is positioned at the centre of the dial. The movement has 21 jewels, 7 adjustments and a patented precision regulator index. (Private collection, A. Simonin)

127 c ▷

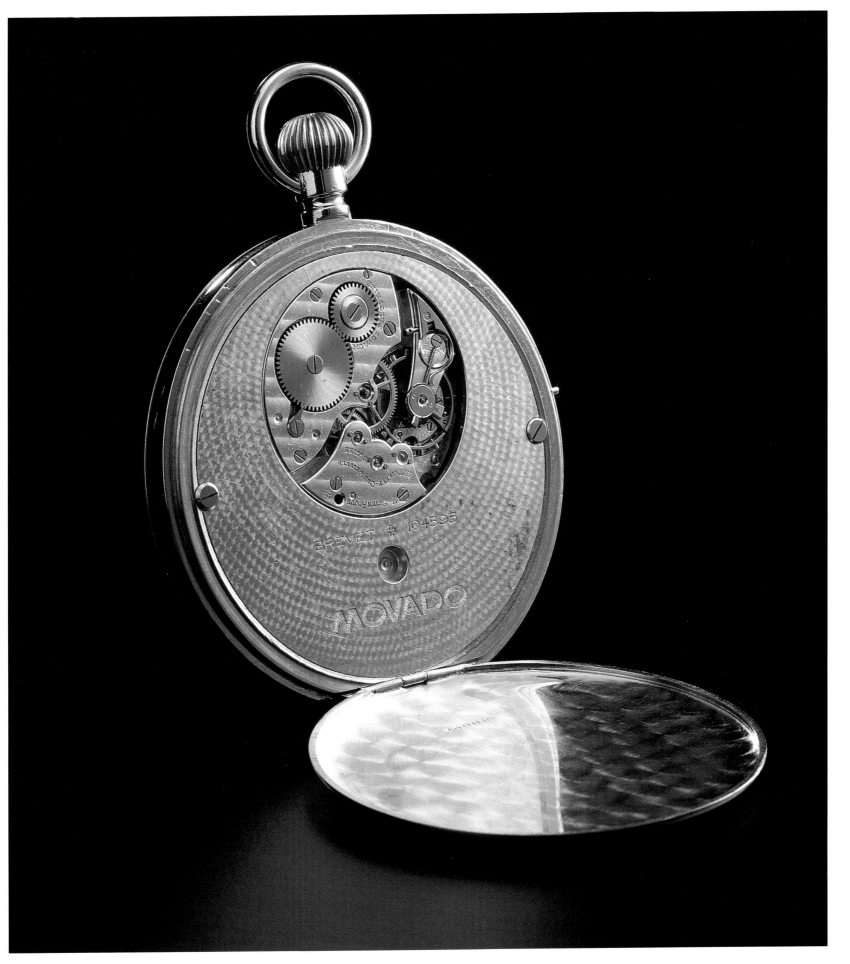

128 a, b
Movado deck watch with an "up" and "down" indicator, set in a walnut box with brass corners, screwed chromium plated bezel, gilded ¾ plate Calibre 355 movement with lever escapement, Guillaume balance, Breguet balance spring, 21 jewels, reference to adjustments "adj. to temp.". Made in 1925.

129
Three diplomas (Neuchâtel
and Teddington) from 1913,
1920 and 1923, extracted
from a 1948 company
brochure.

130 a,b
Movado deck watch with a small
marine chronometer type walnut
box. The complete watch may be
placed in a brass pot with a
screwed bezel, which is supported
by gimbals. The 19‴ Calibre 350
movement with lever escapement
has 21 jewels including 4 gold chatons with screws, a cut compensation
balance, Breguet balance spring
and a fine regulation index. Made
in the 1930's. The watch is the same
as that shown in Illus. 127 but
mounted in a different way.

Developments in the thirties

The world economic crisis of 1931 did not leave Movado untouched. Production had to be drastically reduced until a gradual economic recovery began in 1935. The Ditesheims used these years, in which most factories were either empty or manned by a skeleton staff, as an opportunity to review, to tighten up and re-organise production methods and to develop a broader range of models. The twenties had been a period of searching for new avenues for the wrist watch, with contemporary luxury combined with bizarre and eccentric solutions. Those days were now over. Thoughts now returned towards the sensible, producible and strictly functional. The numerous newly developed watches of this decade were more sober, conventional and affordable.

This was already clear with the first model which appeared on the market in 1931, just before the depression, called the "Curviplan", a name registered as a trademark on 29th May 1931. It was an extended rectangular wrist watch, more than a centimetre shorter than the Polyplan, making its flatter curvature less noticeable. The movement cleverly and compactly almost completely occupied the space available in the case. The recently developed new form calibre 510 with 17 jewels and lightly chamfered edges was employed. Movado protected this form as a registered design in 1930. It was conventional, with the components on a single level. The balance was 9 mm diameter, the same as that of the Polyplan, but the movement did not require the extra constructional work of having the plate angled downwards. The Curviplan existed in four different case and dial variants. In all of them, the winding crown was half set into the case. Four models were available, two in gold and steel, the other two entirely in gold. Owing to the clever use of the casing space, the very elegant Curviplan could be much more domed than most of the contemporary rectangular watches. For example, the short sides of the movement are lightly chamfered when the dial and movement are the same size, thus allowing for the domed dial

while retaining the simple and cost effective construction of a single level movement.

In the same year, 1931, the "Ermeto Bag" appeared. Strictly speaking, it was not a new watch; it was more an accessory, since it used a small Ermeto that had been fitted into the strap of a lady's handbag. More a "design gimmick" than a horological experiment, it is, nevertheless, a highly evocative example of Movado's ceaselessly innovative productivity.

Another of Movado's developments in the depression years was the "Novoplan", which appeared on the market in 1934. It was very similar to the Curviplan, but narrower with a 4.6 mm, 15-jewelled calibre 440 movement of the same length, giving it a somewhat slimmer and more elongated impression. The movement had been developed in about 1930 and was also available as Calibre 443 with centre seconds.

The year 1935 saw the introduction of the first Movado water-resistant wrist watch, the "Acvatic". This unusual spelling of the word is derived from the Latin word "aqua" (water) and was first registered as a trademark on 13th February 1936. Produced in various sizes the Acvatic had a screwed back with a lead gasket and a cork seal for the winding crown. It was developed by the case-making firm of François Borgel of Geneva, which was owned by three brothers called Tauber. They developed many other variously shaped water-resistant cases for Movado during the subsequent decades. Later a two-button chronograph called "Cronacvatic" was developed.

In 1937 the "Cronoplan" was introduced to the market. This watch used the universal calibre 150 MN movement and had a clever new case construction, which enabled long-term time recording (hours and minutes, without seconds) to be undertaken without time-consuming calculations, and to be read directly from the watch. Movado was granted a Swiss patent for this construction, No. 191 277, on 12th November 1936. It consists of two concentric rotatable bezels surrounding the crystal. The inner bezel

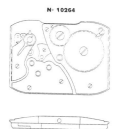

N° 10264

131
Drawing for the protected design of the form movement, Calibre 510, for the "Curviplan", dated 2nd October 1930.

132 a– d
Three Movado gentlemen's wrist watches in bi-coloured gold cases, made in 1925. The upper, rectangular, watch is signed "Chronomètre Movado", Reference No. 0289649 18, with a round 8¾‴, Calibre 760 movement. The middle watch, with tonneau-shaped case, signed "Movado", Reference No. 41570 M0157415, with a round 10¼‴, Calibre 150 MN movement. The lower watch, with tonneau-shaped case and black dial, signed "Movado", has a round 8¾‴ Calibre 730 movement.

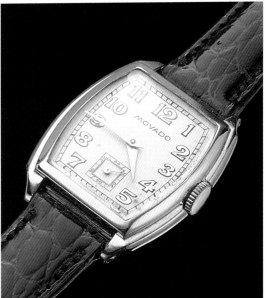

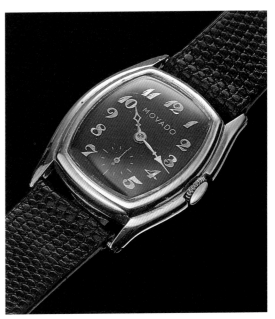

133 a, b
Rectangular Movado "Curviplan" 14 ct gold wrist watch, made in 1935. Silvered dial (restored) signed "Chronometer Movado", with small seconds, Reference No. 41819 527074. Nickel-plated form movement, Calibre 510, for export, signed "Movado Factories", 17 jewels.

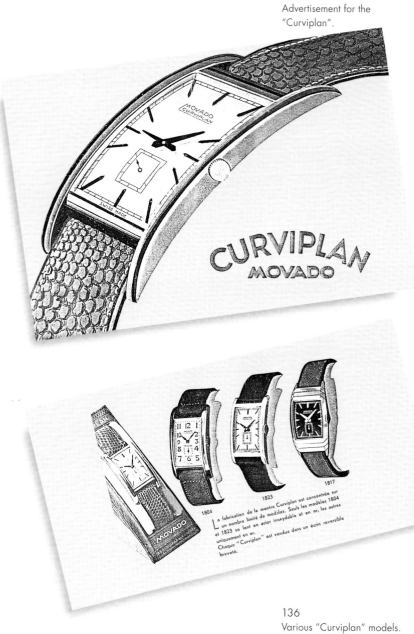

135
Advertisement for the "Curviplan".

134
Two gold cased Movado gentlemen's wrist watches made in 1940. The watch on the left, signed "Movado", with small seconds, extravagantly designed case, Reference No. R 3887 390168. Form movement, Calibre 375, with 15 jewels. On the right, a "Curviplan" signed "Chronometer Movado", with small seconds, Reference No. 41815 526116, Calibre 510 form movement.

136
Various "Curviplan" models.

137
Rectangular, steel-cased
"Novoplan" Movado wrist
watch, made in 1935. The
case has special, broad,
stirrup-shaped lugs. Silvered
dial with small seconds,
signed "Movado", Reference
No. 12224 445534,
Calibre 440 form movement.

138
Rectangular steel gentleman's
Movado wrist watch, made
in 1935. Silvered dial with
separated chapter rings for
hours/minutes and seconds,
signed "E. Gübelin Lucerne",
Reference No. 3880, Calibre
510 form movement with
15 jewels and 4 adjustments.

is divided for hours, the outer for minutes. If one wishes to time an event, the 0 of the inner bezel is placed above the hour hand and the 0 of the outer bezel above the minute hand. At the end of the recording, the elapsed time may be read directly from the numbers indicated by the two hands.

Apart from these series models with their particular and easily remembered names (a knack for which Movado was well known) in the 1930's there were also numerous small series of special models with the time shown in digital form. In this decade, the digital indication of time was very popular. These were watches, generally with rectangular cases, in which rotating discs replaced the hands, as with calendar watches. The time was displayed in small apertures in the case through which the metal discs could be seen. Normal series movements could be employed, where only the motion work required modification. There were models with 2 windows, showing hours and minutes, others with 3 windows boasting an additional seconds indication. Equally popular at this time were combinations of digital and analogue indications, for instance with jumping hours in a window and small minutes and seconds chapters with hands. The renowned movement maker Frédéric Piguet of Le Brassus, nephew and successor of the well-known Louis Elisée Piguet, supplied Movado with 500 examples of a 10½''' movement with such a jumping hour mechanism in the early 1930's.

A rectangular wrist watch of the "Reverso" type, with chronograph and dials on both sides, was made as an experiment in 1939 and is known only since its appearance at a Hong Kong auction in 1990. Indications are that only one example was made and presumably it did not go into production owing to the high manufacturing costs. One of the dials has normal hour and minute chapters with a small seconds below the 12. On the reverse, the dial has two chapter rings, one above the other. The upper one is a 30-minute register and the lower is for the chronograph seconds indication. The chronograph is operated by two slides on the edge of the case. This watch must have an interesting round movement, 8''' (18 mm) diameter with 19 jewels. Such a small chronograph calibre with 30-minute register does not appear in the Movado calibre lists and neither does the movement No. 97 155. There is no doubt that this watch is the only one of its kind.

Doubtless this two-sided watch, made to flip within its own case, was an interesting and unique idea. Unlike the Reverso the function seems designed not to protect the dial but rather to separate the indications of a complicated wrist watch. Such double dials are less useful with normal wrist watches, since the second dial may only be read after unbuckling the watch from the wrist.

Movado already started to develop wrist watches with calendar indications and with chronographs in the 1930's. The experimental model described supposedly belongs to these developments. In 1938 the "Calendograph" and "Chronograph" models came on the market. Both led to a large number of successors, as variants of the models, and we shall follow their traces separately.

139
Rectangular 14 ct gold Movado "Novoplan" wrist watch, made in 1935. The silvered dial with small seconds is signed "Movado", Reference No. 42206 442035, Calibre 440 form movement.

140
Rectangular steel Movado "Curviplan" wrist watch, made in 1931. The silvered dial with small seconds is signed "Movado", Reference No. 11823 515165, Calibre 510 movement.

141
Rectangular gentleman's Movado "Curviplan" wrist watch with silvered dial and Arabic numerals, Calibre 510 movement.

142
Movado "Acvatic" water-resistant wrist watch with 14 ct gold case, made in 1935. The case has a screwed back and distinctively designed lugs for the strap. The two-tone dial with centre seconds is signed "Brock and Co.", Reference No. 41730 0130459, Calibre 150 MN movement.

143
Small round gentleman's Movado wrist watch in a case with a steel bezel and back with a rose gold band, made in 1935. Movable lugs for the strap, silvered dial with centre seconds and highly original numerals, signed "Movado", Reference No. 11727 0184778. Nickel-plated 10¼''', Calibre 157 movement with lever escapement.

144
Round gentleman's Movado wrist watch in a Staybrite steel case with unusual, large case lugs, made in 1935. Two-tone silvered dial with small seconds, signed "Chronomètre Movado", Reference No. 11745 0191046. Calibre 150 MN 10¼''', movement with lever escapement, 15 jewels, 4 adjustments, 3 gold chatons with screws.

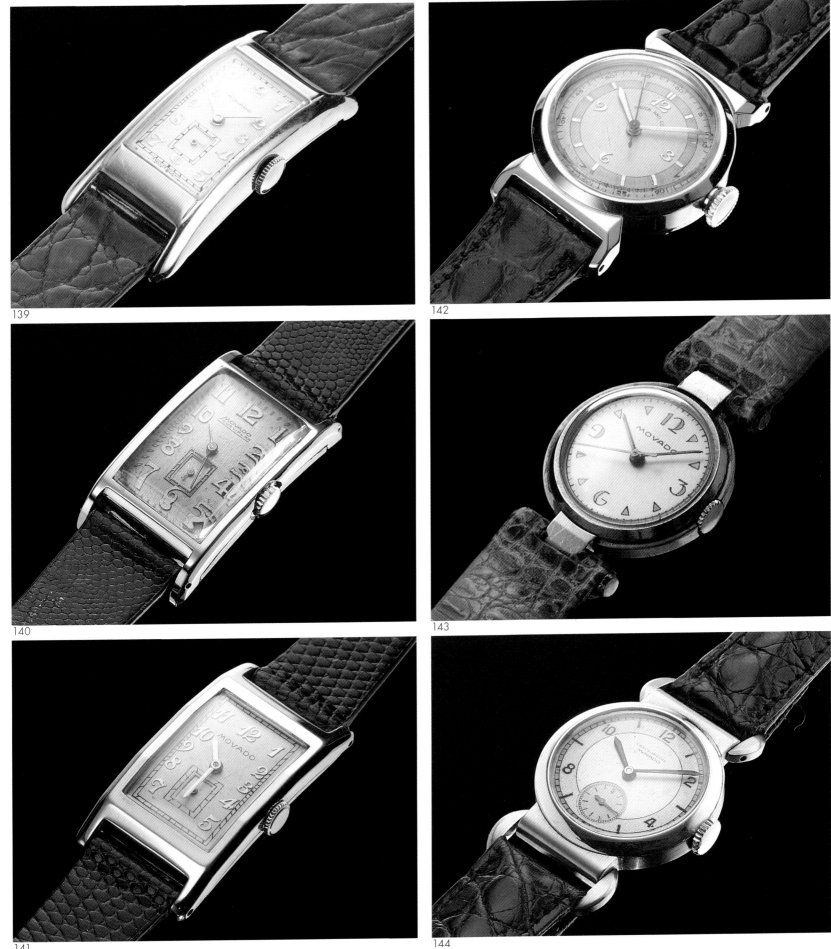

139

140

141

142

143

144

145
Two Movado wrist watches of similar design with steel cases, made in 1940. Left, a two-button chronograph with small seconds and 30-minute recorder plus a pulsimeter scale, signed "Movado". A broad rotatable bezel, divided into 1–12 surrounds the dial, Reference No. 19006 96752, 12''', Calibre 90 movement. Right, the "Cronoplan" model with two external rotatable rings for measuring long periods of time. Silvered dial signed "Movado Cronoplan" with small seconds, Reference No. 11764 157760. Calibre 150 MN movement with 4 adjustments.

146
Patent No. 191 277, dated 15th June 1937, showing the time-measuring system of the "Cronoplan".

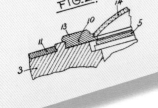

147 a, b
Rectangular, steel-cased, water-resistant Movado wrist watch, made in 1940. The dial, with small seconds, is signed "Movado", Reference No. 456306 12266. The case is stamped with the manufacturer's stamp "FB Patent" (François Borgel, Geneva) and represents one of the first rectangular water-resistant case constructions. Nickel-plated Calibre 440 form movement for export, signed "Movado Factories", with lever escapement and 17 jewels.

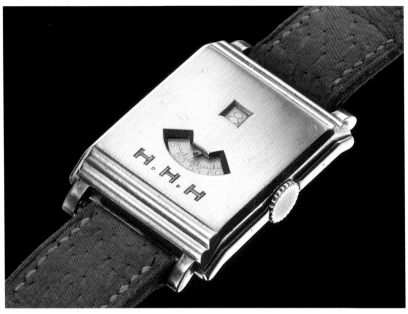

148
Rectangular, silver, Movado digital wrist watch, made in 1931. Signed "Gübelin" on the case back, the case front cover is pierced with two apertures to show the hours and minutes. Reference No. 21701 M0167968. Round, Calibre 150 MN movement with lever escapement.

149
Digital wrist watch with a 14 ct gold water-resistant screwed Acvatic case, made in 1936. The case has very distinctive lugs, Reference No. 166480 41729. Silvered dial with two apertures for the hours and minutes. Nickel-plated 10¼", Calibre 150 MN movement with lever escapement and 17 jewels.

151
Gilded Movado pocket watch with digital time display, made in 1935. The front cover has two apertures for the hour and minute displays. Reference No. 635 304, Calibre 620 movement with lever escapement.

150
Rectangular 14 ct gold digital Movado wrist watch, made in 1930. The case front cover is pierced with two apertures for the hours and minutes. Calibre 510 movement.

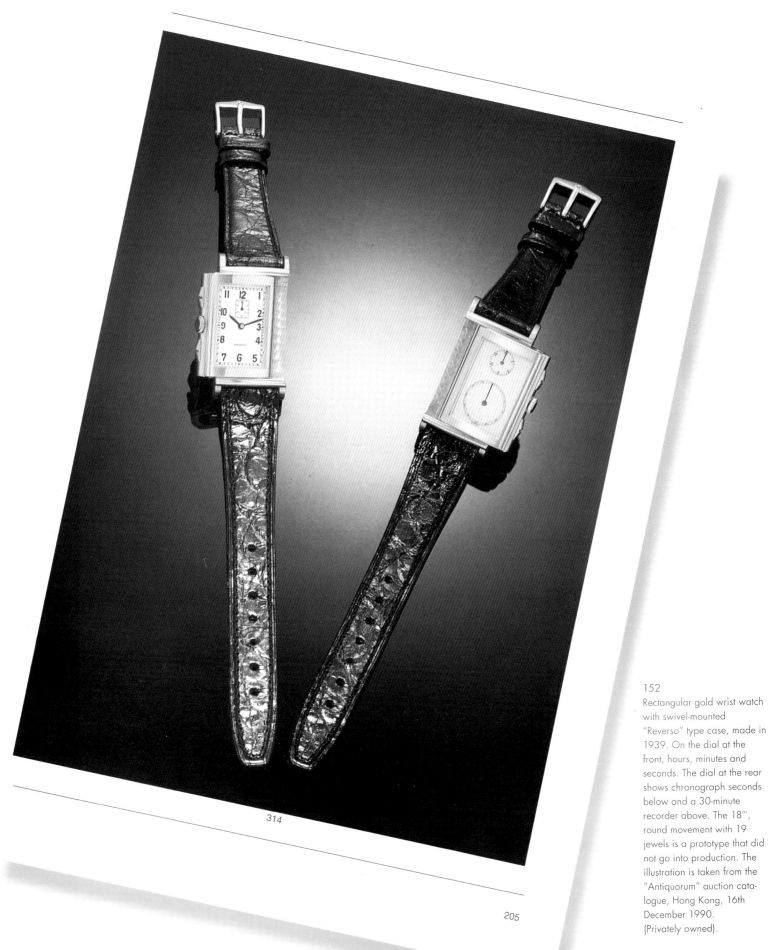

314

205

152
Rectangular gold wrist watch with swivel-mounted "Reverso" type case, made in 1939. On the dial at the front, hours, minutes and seconds. The dial at the rear shows chronograph seconds below and a 30-minute recorder above. The 18''', round movement with 19 jewels is a prototype that did not go into production. The illustration is taken from the "Antiquorum" auction catalogue, Hong Kong, 16th December 1990. (Privately owned).

153 a, b
Rectangular watch set in a brown leather-bound book entitled "Livre d'Heures", made in 1935. The dial is signed "Movado" and the movement is a Calibre 150 MN with lever escapement.

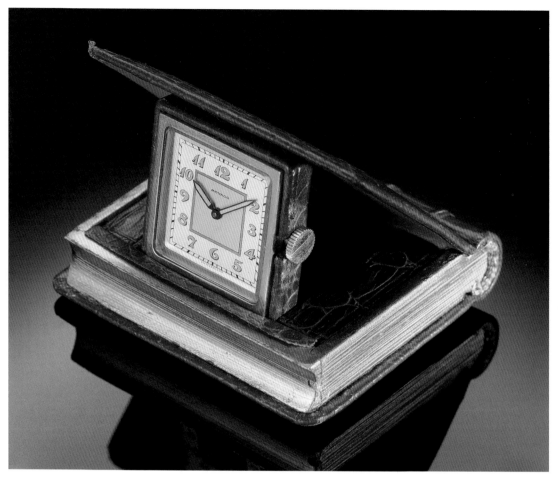

154 a, b
Watch set in a silver belt buckle with coloured enamel inlays, made in 1930. Reference No. 932, Serial No. 0125968. Pressing a catch in the centre portion releases the watch, which then springs out.

155 a, b
Unusual pocket watch of cylindrical form with ball ends, made in 1920. The centre portion is gilded, the ball ends are green enamelled. Pressure on both ends reveals the dial in the centre portion.

156 a, b
Table lighter combined with a watch, made in 1935. The metal lighter case is partially covered with reptile skin leaving one side free to frame the rectangular watch dial. Reference is made to patents in England, Spain and France as well as "made in England", Reference No. 870935. Round, nickel-plated, Calibre 800 M, 19''', movement with modifications, lever escapement, 15 jewels, 4 adjustments. The assembly instructions for the winding mechanism are folded and secured to the movement mounting plate by a screw.

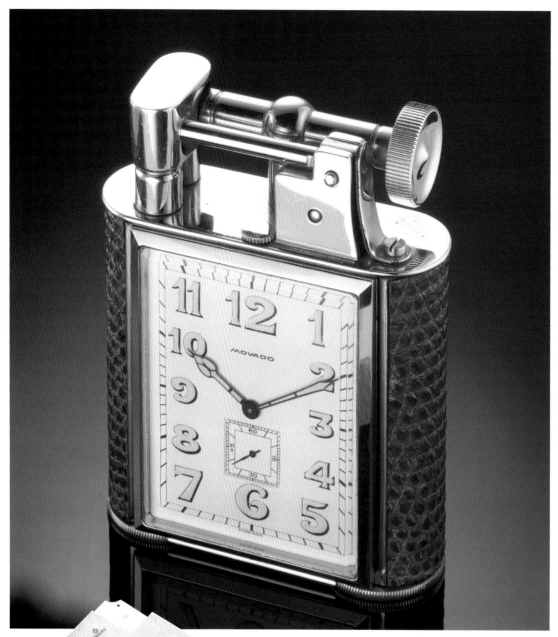

157
Golden ring watch, made in 1950. The dial is signed "Movado", Reference No. 44281, Serial No. 0074463. Nickel-plated Calibre 65 movement with lever escapement and 17 jewels.
▽

159 a, b
Decorated 18 ct gold lady's wrist watch with gold bracelet set with diamonds, made in 1955. The dial is signed "Movado", Reference No. 1475 D18714. Small, round, 7¼‴, Calibre 15 lever escapement movement. The highly original clasp is secured by turning the diamond-set triangle.

158
Four different Movado watches. Upper left, a watch set in a steering wheel with a combination of yellow and rose gold, signed "Tiffany & Co.", with a lever escapement Calibre 50 SP movement. Upper right, a small watch set in a 14 ct gold steering wheel with a lever escapement Calibre 50 SP movement. Lower left, a golden pendant watch, signed "Movado", with a lever escapement Calibre 50 SP movement. Centre right, a golden brooch watch with foliage, signed "Movado", with a lever escapement Calibre 15 movement.

117

Wrist chronographs up to the forties

The list of Movado wrist chronographs is much shorter – this period boasts just two models.

Since 1912 Movado had wrist chronographs with 30-minute registers and a winding crown push piece in production. This placed them among the earliest firms to make such watches and, as already mentioned, Movado presumably sold the first split seconds wrist chronograph under the trade name "Ralco", in 1921. Towards the end of the thirties the two-button chronograph came into fashion. One button serves to start and stop the chronograph hands and the other just returns them to zero. In addition, this model, aimed at "sporty" customers, had either a 30-or 60-minute register and a 12-hour counter. Thus intermediate stopping was possible, i.e. by means of the first button the chronograph hand may be stopped as often as required to let the time be read off and then re-started from this position again. The development of the first Movado two-button chronograph with 60-minute and 12-hour registers was completed in 1938; shortly after the chronograph specialists Universal (1937) and at about the same time as Breitling. The name of the watch was, appropriate to its function, totally prosaic: "Chronograph". The calibre used as basis for the version with only a 60-minute register was the newly developed 12‴ calibre 90 M. The modified calibre 95 M was used for the type with additional 12-hour counter. Here, for the first time, as with the contemporary Calendograph, the supplementary function of the chronograph was a self-contained unit. This made the work easier for a watchmaker carrying out repairs, since the chronograph unit was secured to the basis movement by three prominent blued screws. The chronograph mechanism for the 60-minute register version was developed to order for Movado by the already mentioned firm, Frédéric Piguet of Le Brassus, in 1936, as was the completion of this mechanism with the 12-hour counter for calibre 95 M (in 1939). This was an exception for Movado since they generally developed and made their own movements.

160

Military Movado wrist
chronograph, made in 1940.
The steel case has a screwed
back, the black dial has luminous numerals and hands, no
small seconds, 60-minute
register, signed "Movado".
The 12‴, Calibre 90 M movement has a lever escapement
and 17 jewels. This watch is
presumably a prototype.

161

Movado pilot's wrist watch
with a rotatable bezel and
centre seconds, made in
1935. The steel case is fitted
with a second crown to
locate the rotatable bezel for
measuring short periods of
time. The silvered dial is
signed "Movado", Reference
No. 11796 176688. Nickel-
plated 10½‴, Calibre 155
MN movement with lever
escapement and 15 jewels.

162

Two 18 ct gold cased
Movado wrist chronographs
and a wrist calendar "Celestograph". Left, the calendar
wrist watch, made in 1950.
Centre, a chronograph with
a single push piece in the
winding crown, 30-minute
register and tachometer
scale, made in 1930. Nickel-
plated 13‴ movement with
lever escapement, 17 jewels
and 4 adjustments. Right, a
large chronograph with 60-
minute register, telemeter
and tachometer scales, two
push pieces, small seconds.
Nickel-plated 13‴, movement
with lever escapement and
15 jewels.

One variant of the "Chronograph" was, as already mentioned, the "Cronacvatic" with water-resistant screwed case. Some models had a soft iron capsule to protect them against the influence of magnetism.

In 1941 Movado developed a single button chronograph with the 10¼''' calibre 478, which had a small seconds hand but no counter.

It is actually surprising that Movado did not combine, as did some of their competitors, the chronograph and calendar mechanisms in one watch. This was achieved later in conjunction with Zenith with the renowned "El Primero" model from 1969.

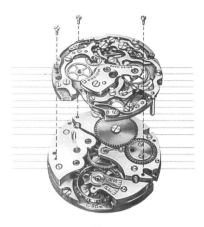

163 a   △
Illustration showing the modular construction of a chronograph.

163 b
Cushion-shaped wrist chronograph, made in 1938. Steel case, silvered dial with a tachometer scale around the outer edge, small seconds, 60- minute register, signed "Movado", Reference No. 19013, movement No. 96961. Nickel-plated 12''', Calibre 90 M movement with lever escapement and 17 jewels.

164 a, b
Round Movado wrist chronograph, made in 1940. Steel,
water-resistant case made to
the FB Patent, silvered dial
with small seconds and
60-minute register, signed
"Movado", Reference No.
A98611 19048. Nickel-
plated 12"', Calibre 90 M
movement with lever escapement, 17 jewels, signed
"Movado Factories".

164 c
Case back showing the
maker's stamp (François
Borgel), upper left and the
"Movado hand symbol",
upper right. In the centre,
"Movado Factories Switzer-
land" and Fabrication Suisse,
and below, Serial No.
A93611 and Case Reference
No. 19048.

165
Wrist chronograph in an 18 ct gold case, made in 1950. Silvered dial with small seconds, 60-minute register, outer pulsometer scale, signed "Movado", Reference No. A99908 R9033. Nickel-plated 12½‴, Calibre 90 M movement with lever escapement and 17 jewels.

166
Wrist chronograph in an 18 ct gold case, made in 1944. Black dial with small seconds, 60-minute register, outer tachometer scale, signed "Movado Nonmagnetic", Reference No. 101907 9002. Nickel-plated 12‴, Calibre 90 M movement with lever escapement.

169 ▷
Platinum-cased wrist chronograph with flexible lugs, made in 1938. Silvered dial with small seconds, 60-minute and 12-hour registers, signed "Movado Non Magnetic". Nickel-plated 12‴, Calibre 95 M movement with lever escapement and 17 jewels. (Privately owned).

167 a, b
Single push-piece Movado chronograph, made in 1945. The case is 18 ct rose gold, silvered dial with small seconds, signed "Movado", Reference No. R4870 A491886. Nickel-plated 10¼‴, Calibre 478 movement with a spacing ring, lever escapement, signed "Movado Factories".

168 a, b
14 ct gold cased wrist chronograph, made in 1941. Silvered dial with small seconds, 60-minute and 12-hour registers, signed "Tiffany & Co", Reference No. 49018 101222. Nickel-plated 12‴, Calibre 95 M movement with lever escapement and 17 jewels.

Wrist calendar watches up to the fifties

The "Calendograph" had the newly developed 12½''' movement, Calibre 475. The novelty of this movement was that the calendar mechanism was a self-contained unit with its own plate (20 mm diameter) secured to the basis movement by two screws. The Calendograph had a simple calendar with the month and day of the week displayed in apertures placed on either side of the central axis and a date indication operated by a centrally placed hand. There was also a model with an indirect centre seconds (Calibre 475 SC) and a slightly smaller version (10½''' diameter), Calibre 479, produced up to 1954.

These were followed by a lady's wrist watch in 1940 with a date hand and the day of the week in an aperture. It was the "Calendette" with the small form Calibre 28. The "Calendoplan Baby" was also a rectangular lady's wrist watch with date and centre seconds. This watch was somewhat smaller overall with the form Calibre 579. In 1945 a version of the Ermeto with the date in an aperture appeared on the market. This was the "Ermeto Calendine", with the small form Calibre 578. A further, and more complicated Ermeto, was the "Calendermeto" of 1948 (Calibre 155) with the day, month and phases of the moon displayed through apertures and a central date hand.

A new epoch was introduced in 1946 with the model "Calendomatic". It had either Calibre 225 or the almost identical 225 M (the M indicated "modification", which in this case was a minimal difference in size). The basis for these calibres was the 220, without the calendar module. The Calendomatic had the full calendar with day and month shown in apertures and the date by a central hand, plus centre seconds and automatic winding, which had existed at Movado since 1945. With these complications, the Calendomatic was not only the first on the market but also the flagship of the Movado fleet in its day. The Calendomatic with Calibres 228/228 A and the marginally smaller Calibres 223/223 A had Incabloc shock absorbers from about 1948.

The "Celestograph" of 1947 may be seen as a modification of the Calendograph. It was

170
Isometric view showing the additional calendar mechanism.

171
French advertisement for the Celestograph, from about 1950.

172 a, b
Movado "Calendograph" wrist watch with calendar indications, made in 1950. Steel case with gold bezel, two-tone silvered dial with apertures for the day of the week and month, the date shown with a hand, centre seconds, signed "Movado", Reference No. D474007 14952. Water-resistant case according to the FB Patent (François Borgel). 10 ¼''', Calibre 475 SC movement with indirect centre seconds, signed "Movado Factories", movement No. 3037, lever escapement and 15 jewels.

173 a, b
Lady's "Calendette" wrist watch with its presentation case, made in 1948. Steel case with rose gold bezel, two-tone dial, signed "Movado Beyer", with outer date chapter ring and aperture for the day of the week, Reference No. 13037 033341. Tonneau-shaped form movement, Calibre 28, signed "Movado Factories", 15 jewels.

equipped with the manually-wound Calibre 473 or 473 SC, with centre seconds, and had, in addition to the full calendar (day, month and date), phases of the moon integrated into the small seconds chapter ring.

As already mentioned, the Calendermeto appeared on the market in 1948, followed by the "Calendolux" in 1950 with Calibre 118. With automatic winding, plus day and date indications, it was a somewhat "slimmed down" Calendomatic. In 1950 came the "Calendoplan", with only a date indication aperture. It was equipped with various movement calibres, such as the manually-wound Calibres 122 and 123 (with centre seconds), 128 and 128 SC (with centre seconds), both with Incabloc shock absorbers. Finally, from 1952, a somewhat more lavish version was introduced with the marginally smaller automatic Calibre 222 with indirect centre seconds as well as the date aperture; this was the "Calendoplan Automatic". At the same time a model with the identical Calibre 224 and a magnifying glass in the crystal above the date window appeared as the "Movascope" model.

This adds up to a confusing variety of models and their variants. We have already referred to the Ditesheims' inventiveness as far as christening their watch models goes. This inventiveness led a reporter from the Horological Journal to make the following sarcastic remark in the May 1950 issue, following a discussion at the 1950 Basel Fair:

"We find the Swiss watch manufacturers have certainly been reading American motor-car advertisements. They begin to apply complicated and almost meaningless names to their products. Without actual condemnation of the idea it must be said that it can go too far. Movado lists a watch it calls Calendomatic and another Tempomatic, whilst yet again there are the Celestograf with a side kick into Calendacvatic, Calendette and the Calendermeto. All this means that the Movado is made with a self-winding movement, also with a calendar and phases-of-the-moon dial. Another model is fitted in a water-resistant case and yet another is in our very old and respected friend "Ermeto", now fitted with a self-winding action. If the manufacturers want the trade to swallow titles like this they will add to the trade's difficulties rather than reduce them."

Today, it is precisely these models that are among the most collectable and nostalgic watches ever produced by Movado.

174
Gentleman's "Calendomatic" wrist watch in a water-resistant 18 ct gold case, made in 1950. Silvered dial, signed "Movado Calendomatic" with apertures for the day of the week and month, the date shown by a hand on an outer chapter ring, centre seconds. The 12''', movement, Calibre 225, has automatic winding.

175
Gentleman's "Calendoplan Automatic" wrist watch with a magnifying lens above the date aperture, marketed as the "Movascope" model, made in 1953. Steel case, silvered dial with centre seconds, signed "Movado Automatic Movascope", date aperture at 3 with a large reading lens in the crystal, Reference No. B 231474 16251. Nickel-plated 12''', Calibre 224 movement, No. 5276, with lever escapement, indirect centre seconds, automatic winding with a swinging weight moving through an arc, 17 jewels.

176
Gentleman's "Celestograph" calendar wrist watch in a 14 ct gold case, made in 1950. Two-tone silvered dial, signed "Movado", with outer date chapter ring, apertures for the month and day of the week, small seconds with integrated moon phase display, Reference No. C 476142 44929. Nickel-plated 10½''', Calibre 473 movement with lever escapement and 15 jewels. This watch served as the model for the calendar wrist watch in the "Collection 1881".

177
Gentleman's "Celestograph" calendar wrist watch in a steel case, made in 1950.

Two-tone silvered dial, signed "Movado", with outer date chapter ring, apertures for the month and day of the week, small seconds with integrated moon phase display, Reference No. C 471337 14923. Nickel-plated 10½''', Calibre 473 movement, signed "Movado Factories", with lever escapement and 15 jewels.

The forties and fifties at Movado

The Second World War left no such clear traces in Movado's output as did the First, with the development of the Military Watch. However, strong sales were no doubt made with the various warring parties in those years, presumably with the robust, water- and dust-resistant wrist watch types, such as the Acvatic and with chronographs. The simple type of military wrist watch with a black dial and strongly contrasting luminous numerals, such as was made by many firms and in large quantities at this time, was not to be found at Movado. There was only a similar sort of prototype chronograph with a black dial, but not corresponding to this type. The good business during the war years does not appear to have been noticeably influenced by the export difficulties with Germany that had arisen due to the Jewish origins of Movado's owners. In any case, with 275 employees, Movado was one of the largest factories in the Canton of Neuchâtel in 1944.

1944 was also the year that the company founder, Achille Ditesheim, the last living member of the founding generation, died. In the extant funeral sermon given by Grand Rabbi Jules Wolff on 1st November 1944, we find the usual commendatory thoughts on the deceased. Beyond the image drawn by Wolff of the harmonious, active life of the deceased, he also spoke of a remarkable personality. Achille, he said, had been a devout member of the Jewish community; a modest and fair person, kind to his employees and friendly to everyone. It is characteristic of the public restraint of the Ditesheim family that his death was made known in the Journal Suisse d'Horlogerie only by a brief announcement, whereas persons of his standing were usually honoured with a detailed obituary.

Further building work to extend the factory in Rue du Parc, La Chaux-de-Fonds, became necessary in 1948. Shortly after its completion, Roger Ditesheim personally decided to have a workshop built in Geneva because skilled workers were in short supply in La Chaux-de-Fonds. The largest expansion of the company also took place in 1948 with branches established on all five continents: in Paris, London, New York, Toronto, Buenos Aires, Cairo, Sydney, Johannesburg and Shanghai. It was Movado's golden age: worldwide representation, responding to stimulation and trends from all corners of the world, a truly cosmopolitan company.

Gérard Ditesheim, who worked for the American branch between 1946 and 1972 and finally became its president, recalls that his father, Roger, was continuously travelling the world at this time. His house in Switzerland, to which jewellers, important customers and well-known personalities were invited, was open and hospitable. One of the most prominent visitors to the factory was Evita, the wife of the Argentinian President, Juan Peron, who was overthrown in 1955. This openness and hospitality was not at all common and certainly not to be taken for granted in Switzerland at that time.

Movado had already applied for a patent for the construction of a self-winding mechanism in 1942. It was for a pendulum type of swinging weight automatic, conceived as a self-contained unit, which was patented on 16th July 1943. But it was not until 1945 that Movado commenced the production of wrist watches with automatic winding. It was achieved with the "Tempomatic" model, which used the Calibres 220/220 M/221 as a basis. This Calibre also abandoned the characteristic cock with three rounded projections, a compulsory movement design up to this time which had become a company trademark. Donald de Carle (Complicated Watches and their Repair, in the Horological Journal 1953/1954) described Calibre 221 as a "simple, clever and well-designed movement". As had already been the case with the Calendograph and Chronograph, this and subsequent automatic movements were made on the module principle, i.e. the automatic winding was a self-contained unit attached to the basis movement by two screws. Technically Movado at first employed the principle of single direction winding, with the pivoted weight swinging between two

179
Steel-cased Movado wrist watch with early automatic winding, made in 1950. The case with prominent and unusual lugs, silvered dial with centre seconds, signed "Movado Tempomatic", Reference No. 16170. Nickel-plated 12¾''', Calibre C 220 M movement with automatic winding by means of a swinging weight, lever escapement, indirect centre seconds, 17 jewels and 2 adjustments.

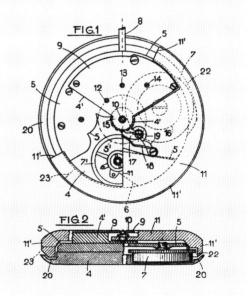

178
Movado advertisement from the 1950's showing the "Tempomatic", the first automatic-winding wrist watch.

180
Patent No. 226 490, dated 15th April 1943, for an automatic-winding mechanism with a swinging weight.

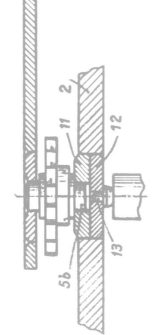

FiG.1

FiG.2

182
Nickel-plated 11¾‴, Calibre 538 movement with lever escapement and somewhat increased vibrations (21 600 vibs./hour), signed "Movado Factories", with direct centre seconds and automatic winding from a central rotor, 28 jewels.

183
Movement signed "Movado Factories", 11¼‴, Calibre 395 E, with lever escapement, direct centre seconds, automatic winding from a central rotor, 17 jewels.

181
Patent No. 279 956, dated 31st December 1951, for the special construction of a flat movement with automatic winding using a common jewel bearing to pivot the winding weight and the centre pinion (Calibre 115 "Futuramic").

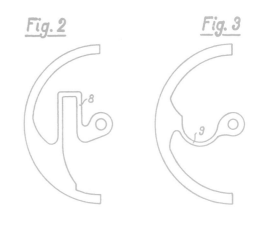

Fig. 2

Fig. 3

184
Patent No. 281 490, dated 15th March 1952, for an s-shaped flexible mounting for the winding weight of the automatic Calibre 115 "Futuramic".

◁ 185 a
Nickel-plated 12½‴, Calibre 230 movement (Universal Calibre 66) with lever escapement, eccentric micro-rotor, 25 jewels, mono-metallic balance, signed "Movado Factories".

185 b
18 ct gold-cased automatic-winding wrist watch, made in 1965. Silvered dial without seconds, signed "Movado Automatic".

186 a, b
14 ct gold-cased Movado wrist chronometer, made in 1952. Silvered dial with small seconds, signed "Chronomètre Movado 21 jewels Automatic". Nickel-plated 11¾''', Calibre 116 movement for export, movement No. 8005, signed "Movado Factories", lever escapement, automatic winding from a swinging weight, 21 jewels, adjusted for temperature and 5 positions, Glucydur balance.

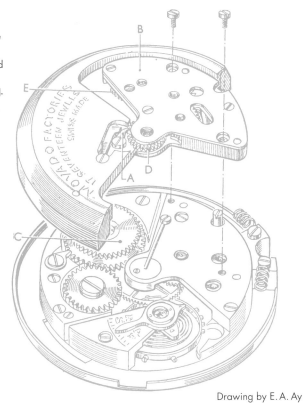

Drawing by E. A. Ay from a book by D. De Carle.

187
Patent No. 281 490, date 15th March 1952, for an s-shaped flexible mounting for the winding weight of automatic Calibre 115 "Futuramic".

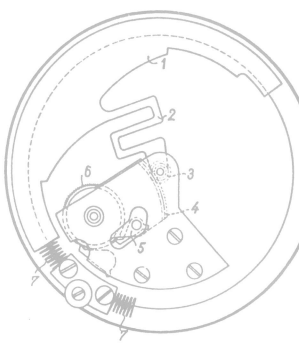

buffer springs, although the more modern and future-oriented version with a rotor winding in both directions already existed by 1942. All movement calibres of the Tempomatic had a centre seconds hand. Some calendar wrist watches, using Calibre 220 as a basis, were also equipped with automatic winding, such as the Calendomatic of 1946 with Calibre 225 and the Calendoplan, developed further in 1952, using Calibre 222.

A further development of automatic winding was the 11½‴ Calibre 115 "Futuramic" that came out in 1950. This was likewise a single direction winding system with a swinging weight. The latter had a thin s-shaped spring-like extension to pivot the weight at the centre of the watch and give it flexibility in addition to that of the spring buffers at the sides. The patent of 15th March 1952 records two alternative forms of the flexible extension. Incidentally, this patent is separate from a second one, dated 31st December 1951, for the special bearings of the swinging weight of this Calibre. This movement was noticeably flatter (with an overall height of 4.45 mm) than Calibre 220 (5.7 mm), fitting in with the 1950's trend toward ever flatter wrist watches. Movado advertised this calibre as the flattest yet on the market. This low overall height was achieved by employing a particularly narrow heavy metal weight which shared a single jewel bearing with the centre wheel. Moreover, there was no central seconds. A further development was Calibre 118, the "Calendolux" of about 1950, which had the addition of day of the week and date displays thus adding 1 mm to the thickness.

The first additional automatic mechanisms with a rotor were introduced at Movado in 1956; with the 11¼‴ Calibre 431 A with centre seconds for the "Kingmatic" model which was to become a well-known and long-lived series. At the same time the ladies' model, the "Queenmatic", appeared with the 7¼‴ automatic with rotor, Calibre 165. This was replaced by the somewhat flatter Calibre 421/423/425/427 with a slightly faster beating balance with 19 800 vibrations per hour. Two of these last-mentioned calibres, 423 and 427, had indirect centre seconds and the other two both had small seconds. Also the rotor automatics were soon combined with calendar mechanisms; in 1959 Calibre 438, which was a further development of Calibre 431 A, with the date shown in an aperture for the "Kingmatic Calendar". Calibre 438 was in turn developed to

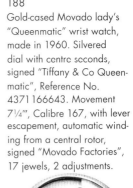

188
Gold-cased Movado lady's "Queenmatic" wrist watch, made in 1960. Silvered dial with centre seconds, signed "Tiffany & Co Queenmatic", Reference No. 4371166643. Movement 7¼‴, Calibre 167, with lever escapement, automatic winding from a central rotor, signed "Movado Factories", 17 jewels, 2 adjustments.

189
Gentleman's 18 ct gold-cased wrist watch with solid gold dial, made in 1955. The dial has an engine-turned inner square panel, small seconds, signed "Movado", Reference No. R8463 A2133995. Nickel-plated 11¾‴, Calibre 115 movement, No. 8405, with lever escapement, automatic winding from a swinging weight, 19 jewels.

190
Movado "Kingmatic" wrist watch, made in 1965. Gold-capped steel case, silvered dial, signed "Movado Kingmatic", centre seconds, Reference No. 19110. Automatic winding movement with central rotor, 11¼‴, Calibre 395 E, lever escapement, direct centre seconds, signed "Movado Factories", 17 jewels. See Illus. 183 for a view of the movement.

191

Movado 14 ct gold-cased automatic winding watch, made in 1950. Silvered dial with small seconds, signed "Movado Automatic", Reference No. 211720 48451.

The movement is an 11¾‴, Calibre 115 with lever escapement, swinging weight automatic winding, 17 jewels.

192
Movado advertisement from 1951 for the flat automatic, Calibre 331.

the almost identical Calibre 531. It was without the calendar module but the vibrations were increased to 21 600 per hour in 1960 to improve the rates of the "Kingmatic" model. Greater precision was the reason for the simultaneously introduced Calibres 536 and 538, the latter with the date shown in an aperture, both with direct centre seconds. They were mainly reserved for wrist chronometers, in particular the "Kingmatic Chronomètre" as well as the "Kingmatic Calendoplan" with Calibre 538.

In the 1960's the drive to achieve yet flatter automatic movements continued. The Calibre group 380 of 1966 for the "Kingmatic S" was only 4 mm high and had a rotor supported by a ball bearing. Co-operation with Universal led to the adoption of the Universal Calibre 66 which was only 2.5 mm high with an offset micro-rotor. This was designated No. 230 at Movado and was employed in the ultra-flat gent's wrist watch "Gentleman".

So much for the account of the automatic movements with their numerous lines of development – and names, which one may either find amusing, or simply accept.

Shortly after 1950 the world time watch "Polygraph" came on the market. It was patented by Movado on 30th November 1951. The movement was Calibre 129, which was Calibre 125 with the addition of a world-time mechanism to the basic calibre. It was based on the mechanism developed by the Genevan watchmaker Louis Cottier, and first constructed on his principle by Patek Philippe in 1937 as the "Heure universelle". The dial consists of a fixed 12-hour chapter ring around which there is a narrow 24-hour ring, and then, externally, a manually rotatable bezel with a milled edge. The bezel is engraved with the names of cities in various time zones. In addition there is a central 24-hour hand which indicates the time on the 24-hour chapter ring. If one wishes to read off the time in a particular zone, the name of a city in that zone is placed opposite the 24-hour hand (by means of the rotatable bezel) and the time in the corresponding zone may then be read on the 24-hour chapter ring.

Likewise in 1950, a wrist watch with a 24-hour dial, called the "Astronic", was placed on the market. This watch used the very flat manually wound Calibre 260 SC (with centre seconds). Later the Astronic was equipped with the even flatter Calibre 346 movement, which had an overall height of 3.4 mm.

193

Steel-cased World Time wrist watch, made in 1950. Multi-coloured dial with a rotatable outer bezel, small seconds, an additional central 24-hour hand, signed "Movado". Model designation "Polygraph", Reference No. 18150 1167395. Nickel-plated 12½''' Calibre 129 movement for export, with lever escapement, signed "Movado Factories", 17 jewels.

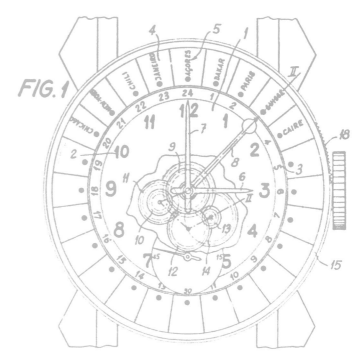

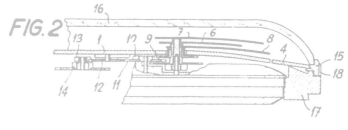

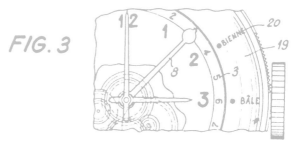

194
Patent No. 279 370, dated 30th November 1951, for the mechanism of the World Time "Polygraph".

195
Advertisement for the 1957 "Calendoscope" model.

Now some additional comments on the "Kingmatic", which has already been mentioned several times. This very successful model group was produced in large numbers right into the 1970's. In the 1950's and 60's this watch may be regarded as the standard Movado model. It came onto the market in 1956 together with the first rotor automatic calibre. By 1960, the Kingmatic was equipped with the somewhat faster vibrating and more precise Calibre 531 movement; and the "Kingmatic Chronomètre" with Calibres 536 and 538, by the same year. After 1966, the classic "Kingmatic S" was flatter, as already stated, and had a ball-bearing rotor. In the version showing the date through a window, the former could be corrected by means of the crown without turning the hands. The Calibre group 395 A-F was also used for the Kingmatic S. From 1968 on, the "Kingmatic HS 360", the most modern and last version, used the fast train Calibres 404 to 409 with 36 000 vibrations per hour and various calendar variants.

Some impressive advertising demonstrations were also undertaken with the Kingmatic. In 1960 a "Kingmatic Sub Sea" was attached to the outside of the hull of an ocean liner in such a way that it was permanently submerged throughout an Atlantic crossing; the watch withstood this test, its function was completely unaffected. A few years later an "Opération Précision" was carried out in which two Air France Captains were each given a Kingmatic, exactly synchronised to each other. One of them flew the route Paris-Los Angeles-Paris while the other flew Paris-Tokyo-Paris, and after their return 7 days later the two watches differed by only half a second.

196
Gold-cased Movado wrist watch with a map of Saudi Arabia on the dial, made in 1956. The dial with an outer gilded chapter ring surrounding a multi-coloured map of Saudi Arabia, signed "Movado", without seconds, Reference No, R 8463 212011. Nickel-plated 11¾''' Calibre 115 "Futuramic" movement with lever escapement, automatic winding with swinging weight, 17 jewels.

197
Luxury "Kingmatic" gentleman's 18 ct white gold wrist watch with a white gold link bracelet and diamond-set solid white gold dial, made in 1966. The white gold dial has an engine-turned inner square panel, centre seconds, date in an aperture at 3, signed "Movado Kingmatic", all numerals represented by either rectangular or square-cut diamonds, Reference No. 538 271 583. The screwed water-resistant back is inscribed "Kingmatic Sub-Sea". Nickel-plated 11¾''' Calibre 538 movement with lever escapement and somewhat increased vibrations (21 600 vibs./hour), signed "Movado Factories", direct centre seconds and automatic winding from a central rotor, 28 jewels (See Illus. 182).

198 a, b
Steel-cased Movado wrist watch, made in 1945. The rose gold capped case is water-resistant with a screwed back. Two-toned coloured dial, signed "Tiffany & Co", Reference No. 13308 142349. Nickel-plated 8¾''', Calibre 105 SC.

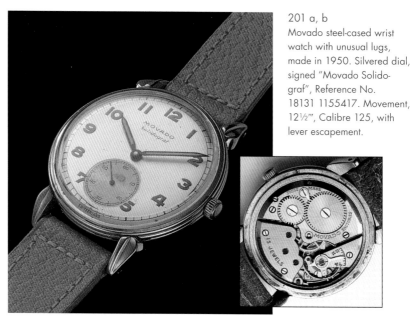

201 a, b
Movado steel-cased wrist watch with unusual lugs, made in 1950. Silvered dial, signed "Movado Solidograf", Reference No. 18131 1155417. Movement, 12½''', Calibre 125, with lever escapement.

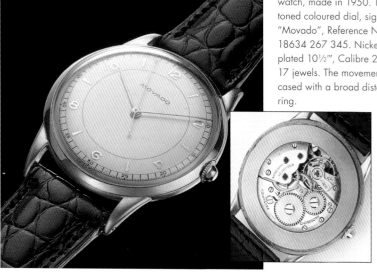
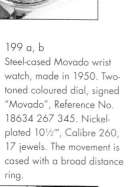

199 a, b
Steel-cased Movado wrist watch, made in 1950. Two-toned coloured dial, signed "Movado", Reference No. 18634 267 345. Nickel-plated 10½''', Calibre 260, 17 jewels. The movement is cased with a broad distance ring.

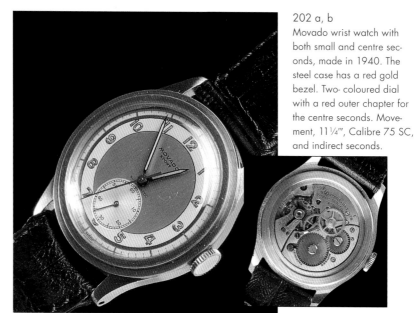

202 a, b
Movado wrist watch with both small and centre seconds, made in 1940. The steel case has a red gold bezel. Two-coloured dial with a red outer chapter for the centre seconds. Movement, 11¼''', Calibre 75 SC, and indirect seconds.

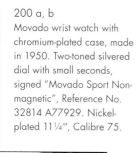

200 a, b
Movado wrist watch with chromium-plated case, made in 1950. Two-toned silvered dial with small seconds, signed "Movado Sport Non-magnetic", Reference No. 32814 A77929. Nickel-plated 11¼''', Calibre 75.

203
Flat, steel-cased, wrist watch with a 24 hour dial, Model "Astronic", made in 1945. Multi-coloured dial with 24-hour chapter ring, centre seconds, signed "Movado Astronic", Reference No. 18901 265045. Movement, 10½''', Calibre 260 with lever escapement and indirect seconds, 17 jewels.

204
Steel-cased Movado wrist
watch, made in 1950. Dial
with small seconds and 6
cabochon rubies, signed
"Movado", Reference No.
14854 477583, with origi-
nal presentation case. Movement, 10¼''', Calibre 470
with lever escapement,
15 jewels.

205
Movado wrist watch cased in
18 ct gold, made in 1940.
Delicate rose gold case,
silvered dial with small
seconds, signed "Movado",
Reference No.
R 4859 B 489631. Move-
ment, 10¼''', Calibre 470
with lever escapement ,
15 jewels.

206 a, b
Lady's 14 ct gold-cased wrist watch, made in 1950. Silvered dial without seconds, signed "Movado", Reference No. 18088 41561. Nickel-plated 7¼''', Calibre 15 movement with lever escapement, signed "Movado Factories", for export, 17 jewels.

208
Lady's wrist watch with integral bi-coloured (yellow and white) gold bangle, made in 1960. The back of the case is inscribed "Made in France". Square dial without seconds, signed "Movado". Form movement, Calibre 48 with lever escapement, 17 jewels.

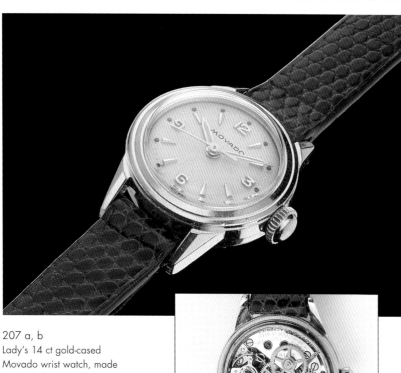

207 a, b
Lady's 14 ct gold-cased Movado wrist watch, made in 1956. Silvered dial with centre seconds, signed "Movado", Reference No. 41751 21909. Nickel-plated 7¼''', Calibre 19 SC movement with lever escapement, indirect centre seconds, movement No. 1195, signed "Movado Factories", 17 jewels, 2 adjustments.

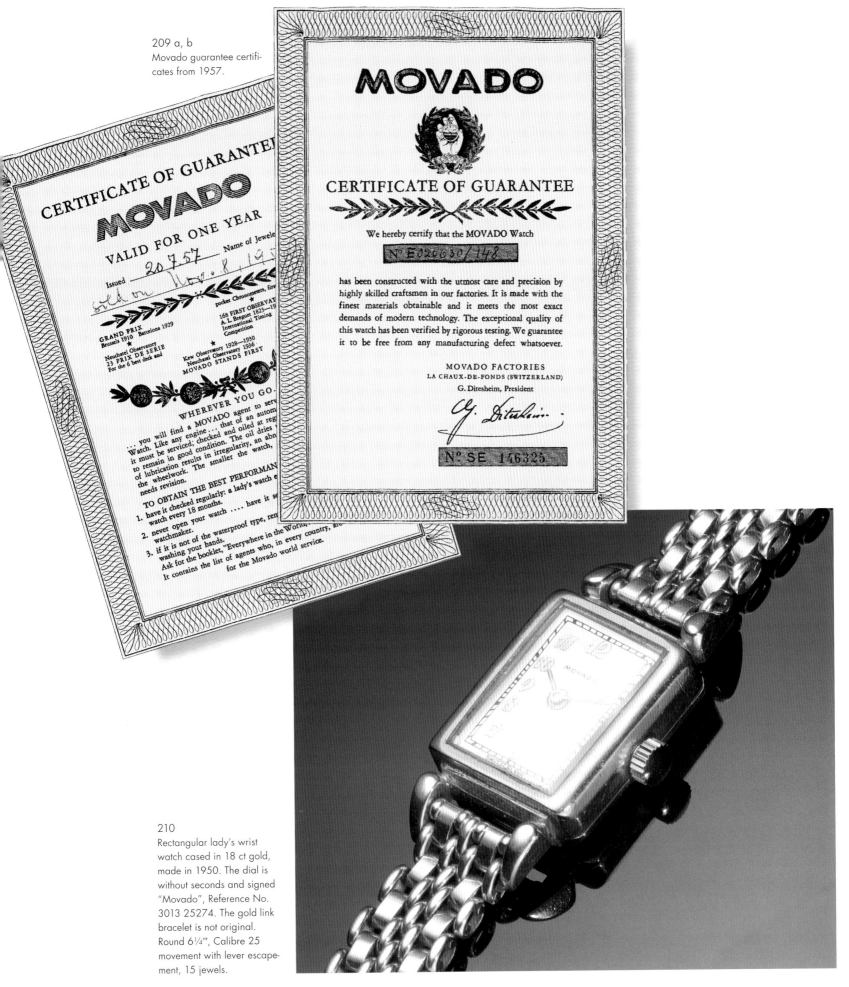

209 a, b
Movado guarantee certificates from 1957.

MOVADO

CERTIFICATE OF GUARANTEE

We hereby certify that the MOVADO Watch

N° E020030/148

has been constructed with the utmost care and precision by highly skilled craftsmen in our factories. It is made with the finest materials obtainable and it meets the most exact demands of modern technology. The exceptional quality of this watch has been verified by rigorous testing. We guarantee it to be free from any manufacturing defect whatsoever.

MOVADO FACTORIES
LA CHAUX-DE-FONDS (SWITZERLAND)

G. Ditesheim, President

N° SE 146325

CERTIFICATE OF GUARANTEE

MOVADO

VALID FOR ONE YEAR

Issued 20757 Name of Jeweler

sold on Nov. 8, 19

pocket Chronometers, firs

GRAND PRIX
Brussels 1910 Barcelona 1929

168 FIRST OBSERVAT
A. L. Breguet 1823—19
International Timing
Competition

Neuchatel Observatory
23 PRIX DE SERIE
For the 6 best deck and

Kew Observatory 1928—1930
Neuchatel Observatory 1936
MOVADO STANDS FIRST

WHEREVER YOU GO...

... you will find a MOVADO agent to serv
Watch. Like any engine ... that of an autom
it must be serviced; checked and oiled at reg
to remain in good condition. The oil dries
of lubrication results in irregularity, an abn
the wheelwork. The smaller the watch,
needs revision.

TO OBTAIN THE BEST PERFORMAN

1. have it checked regularly: a lady's watch e
   watch every 18 months.
2. never open your watch .... have it s
   watchmaker.
3. if it is not of the waterproof type, ren
   washing your hands.
   Ask for the booklet, "Everywhere in the World,
   It contains the list of agents who, in every country, ar
   for the Movado world service.

210
Rectangular lady's wrist watch cased in 18 ct gold, made in 1950. The dial is without seconds and signed "Movado", Reference No. 3013 25274. The gold link bracelet is not original. Round 6¼''', Calibre 25 movement with lever escapement, 15 jewels.

211
Movado advertisement from the 1950's.

212 a, b
Rectangular lady's wrist watch cased in 14 ct gold with unusual decorated lugs, each set with a trapezoidally-cut topaz and diamonds, made in 1955. The dial is without seconds and signed "Movado", Reference No. 401 70615. Nickel-plated form movement, Calibre 65, for export, signed "Movado Factories", with lever escapement, 17 jewels. The watch is in its original presentation case.

1950 – 1969: Wrist chronometers

The regular annual participation of Movado in the *Swiss Institutes for Official Watch Timekeeping Tests (B.O.)* also falls within the period 1950 – 1969. In these tests, normal series wrist watches, destined for sale, were checked to find out if they conformed to the minimum prescribed rates in 5 different positions and at 3 temperatures. After passing these tests, the watches received the designation "Chronometer" with a confirmatory certificate. These trials commenced in 1927. Up until 1951 the companies had the choice of testing the watches themselves (according to the compulsory test procedure) or giving them to one of the 7 Testing Offices to test. From 1952 on, only the chronometer trials conducted at one of the 7 Testing Offices were permitted.

Between 1927 and 1950, Movado presented individual watches for testing only sporadically, exclusively at the local office in La Chaux-de-Fonds (1 pocket watch and 7 wrist watches in 1927, a single wrist watch in 1934 and a pocket watch in 1938). Presumably Movado had taken advantage of the possibility to test watches (in unknown quantities) to chronometer accuracy in-house at this time, since there are numerous known Movado wrist chronometers, particularly from the 1930's and 40's, e.g. with Calibre 150 MN, or Curviplan models with Calibre 510.

Each year from 1950 until 1969, Movado regularly submitted a quantity of wrist watches to the B.O. in La Chaux-de-Fonds. The quantities were always very low and varied between 100 – 300 per annum. Very few in comparison with the market leaders submitting for testing (per company) up to 20,000 watches in 1950 and about 200,000 in 1969. Only in 1966 and 1967 was there a noticeable increase in Movado's submissions, to 366 and 842 watches respectively. In 1968 and 1969, Movado submitted for testing a quantity of wrist watches with complications (1968 : 405 watches, 1969 : 193 watches). These were, according to the rules, watches with calendar or chronograph, or both combined, for which higher limits applied. The fusion with Zenith

brought Movado's chronometer activities to a close. All the Movado wrist watches designated in the trials as chronometers since 1950 achieved the better of the two possible categories with lower limits (designated a.m. = avec mention = pass with distinction. The other, lower level category, s.m. = sans mention = pass without distinction).

Which movement calibres were employed for B.O. tested wrist chronometers after 1950? First of all the automatic Calibre 116 which came out in 1950 as a further development of the flat 17-jewel Calibre 115, but now with 21 functional jewels. Perhaps it was the qualities of this calibre that promoted the subsequent regular participation in the trials. Also the even flatter (3.9 mm) manually-wound Calibre 126, which also appeared in 1950, was largely reserved for chronometers. Frequently one also comes across Kingmatic models with Calibres 536/538 that are chronometers. From 1966, the Calibre group 380/388/389 was used for the Kingmatic S as a chronometer. Possibly this group was the reason for the increase in submissions in the years 1966 and 1967: to test more extensively in practice the rates of the newly developed calibres. Finally, after 1968 there was also the chronometer Kingmatic HS 360 with the new fast train Calibres 408/409. Both had instantaneously changing date indications shown in an aperture, and were perhaps the complicated wrist chronometers of the years 1968 and 1969 that have already been mentioned. Furthermore it cannot be ruled out that other movement calibres, after having been taken from a series and undergoing special adjusting, were also used for chronometers.

In 1945, the two Swiss Observatories at Neuchâtel and Geneva opened the chronometer trials and competitions to wrist watches as well, laying down special regulations for them. Contrary to what has been described hitherto, the wrist watches tested were not pieces intended for sale (this only occurred in a very few exceptional cases) but were particularly carefully pre-

213 a, b
Round 18 ct gold-cased wrist chronometer, Model No. 8103, Serial No. 125 646, made in 1950. Silvered dial with small seconds, signed "Chronomètre Movado, 21 jewels, Fab. Suisse". Manually-wound Calibre 126 movement in a special chronometer execution (based on Calibre 125) with a swan-neck precision regulator index, Glucydur balance with a Breguet balance spring, end stones set in a plate for third wheel, fourth wheel and escape wheel pivots, jewels set in gold chatons. Special engraving: "Adjusted to Temperature and five (5) Positions, 21 Twentyone Jewels, Swiss Made, Movado Factories".

214 a, b
Movado 18 ct gold cased wrist chronometer, made post 1966. Silvered dial with centre seconds, date in an aperture at 3, signed "Chronomètre Movado Kingmatic S", Reference No. 388 214 530. Nickel-plated 11¼''', Calibre 388 movement, No. 9979, signed "Movado Factories", with lever escapement, direct centre seconds, automatic winding from a central rotor, 28 jewels, adjusted for temperature and 5 positions.

THE HEIGHT OF PRECISION

The most recent Timing Competitions conducted at the official Swiss Neuchatel Observatory demonstrates Movado's leadership in the timekeepers pursuit of perfection. For the third consecutive year, Movado has swept the field.

FIRST in the grand list, for performance accuracy of any single wrist chronometer. First in the Series Prix, for performance accuracy for wrist chronometers (breaking all previous records).

Close formation flying at supersonic speeds calls for super-human precision and timing. But that precision is crude compared to the accuracy of a Movado chronometer.

"KINGMATIC" Chronometer, self winding, water resistant, 28 jewels, with certificate from "The Swiss Bureau for the official control of watch performances": in 18 K. Gold, $350; in Steel and 14 K. Gold, $200. Other Movado self winding watches, 17 jewels, from $85. Fed. Tax Incl.

for those whose moments are precious

MOVADO

Sold and serviced by leading jewelers all over the world. For jeweler nearest you write: MOVADO, 610 Fifth Ave., N. Y. In Canada: 44 King St., West Toronto

215
1959 advertisement for the "Kingmatic Chronometer" model.

pared and adjusted individual examples, not included in the series, but entirely based on series calibres. The reason for this particularly careful manufacture, precision improvements and adjustment of the watches, was that the trials, and especially the competitions, demanded very much better rates than those required for the B.O. tests.

Movado did not enter the trials and competitions immediately in 1945, but only in 1950, at the same time, therefore, as the regular participation in the B.O. tests commenced. With the three wrist watches entered in 1950, Nos. 330 003, 330 004 and 330 007, prepared by the renowned adjuster Werner-Albert Dubois who had been working for Movado since the 1930's, no prizes were won, only 34th, 39th and 42nd places were achieved.

To be able to judge such placings correctly it is necessary to understand the evaluation system used by Neuchâtel Observatory. The results obtained during the 45 day trials were expressed as a number of points given to the watch in question. These points were given in such a way that the number for an error-free watch would be zero. The better a watch performed, the lower the measured rate deviations would be and the lower the number of points. Three prize groups were formed among the best competition watches. A first prize was awarded to wrist watches receiving between 0 and 8.5 points. Second prizes were given for 8.5 to 10 points and third prizes for 10 to 12 points. The watches were placed within the prize groups according to their numbers of points. For example, a first prize and the 1st place (as for No. 222 in 1958) signified the best watch of the entire competition with the lowest number of points of the whole year. Watches scoring under all the set limits but with more than 12 points had passed the chronometer trials and were permitted to be designated "Observatory Chronometers" (Chronomètre d'Observatoire) but did not receive awards. The three Movado wrist watches which did not receive awards in 1950 had 13.4, 15.7, and 16.6 points. The prize grouping was abolished in 1963. There were then only award winning watches, with between 0 and 7.5 points or watches without awards with more than 7.5 points.

Obviously the 1950 results were disappointing. Perhaps the veteran adjuster Dubois, specialised in pocket chronometers, was not the right man for wrist watches. It was only in 1956

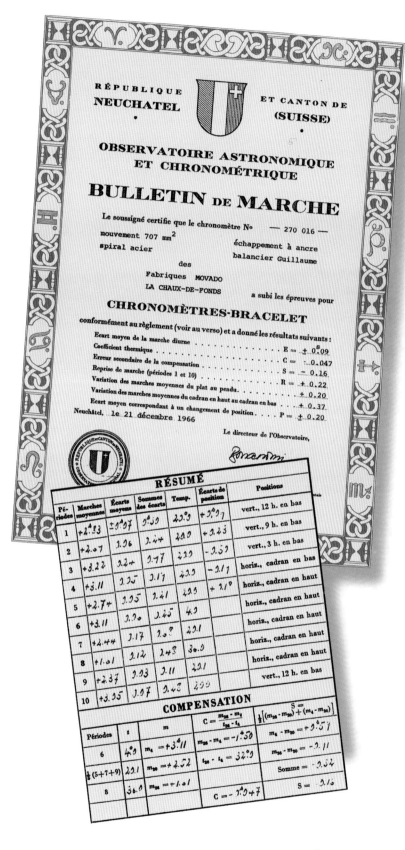

216 a, b
Rating certificate from Neuchâtel Observatory for wrist chronometer No. 270 016, made in 1966.

that a new attempt was made with more watches being submitted. These had been adjusted in the same year by Georges Sautebin, who was specialised in wrist watch movements and was also a master adjuster and teacher at the School of Horology in La Chaux-de-Fonds. Movado won a 1st prize at the first attempt with chronometer No. 210, two 3rd prizes as well as a series prize for the four best wrist chronometers.

These successes were continuously repeated up until 1966 when the wrist chronometer trials were finally abandoned. According to the lists we may deduce that Movado entered between 3 and 13 chronometers annually, and after 1950 most of them won prizes. Three movement numbering systems were used and (presumably only) two movement calibres.

The movement numbers for the year 1950 were of six figures, of which the first three were 330. If one removes these latter figures one presumably has the chronometers Nos. 3, 4, and 7. After Sautebin was engaged as adjuster, the numbering system was changed: from then on the movement numbers were three-figured, commencing with 200. This system was retained until 1963. They commence with No. 201 and end with No. 237. Three movement numbers are missing from the Observatory lists so that only 33 chronometers may be ascertained. However, Movado presumably prepared a total of 37 watches, within this numbering system, for the trials. In the period up to 1963 Movado used the largest permitted round movement calibre of 13''' size (30 mm). This movement was based on the Ebauche Company Peseux Calibre 260 (not to be confused with Movado Calibre 260) and was known as Calibre 30 P at Movado. The Peseux Calibre was also used by other companies such as Ulysse Nardin and Longines for their competition chronometers. (Movado apparently increased the balance vibrations from 18 000 to 21 600 vibrations per hour.*)

In 1961 Georges Sautebin was replaced by the adjuster Henri Guye. At the same time as a new prize group system was introduced in 1964, Movado returned to six-digit movement numbers with 270 as the first three digits. If we delete these three digits referring to Calibre 270 the series commences with No. 14 (270 014) and ends in 1966 with No. 43; there were also 29 observatory chronometers within this numbering system. Together with the watches from the other two numbering systems, it would appear that Movado had a total of 69 wrist chronometers which were entered in the trials.

With the introduction of Calibre 270 in 1964, the shape of the movements was also changed: round movements were no longer used and were replaced by rectangular ones with the maximum permitted movement surface area of 707 sq. mm. The rectangular Calibre 270, no examples of which have so far come to light, had a fast beating balance with 36 000 vibrations per hour and the Clinergic 21 escapement. However, it apparently did not achieve the good rates that were anticipated.

Movado only participated in the Geneva Observatory competitions on two occasions: in 1962, the six 30 mm round movements Nos. 215, 221, 227, 230, 231 and 237 won a manufacturer's series prize; and in 1966, two placings were achieved with the rectangular Calibre Nos. 270 029 and 270 034.

The majority of the chronometers were submitted for the competitions on numerous occasions, predominantly twice, sometimes up to four times; this was generally the case. Some watches participated twice in the same year, namely in Neuchâtel first, followed by Geneva, or the other way round (Nos. 221, 227, 230 and 270 034). Chronometers that were no longer used for competitions were either retained by the company or the adjuster. Only very rarely was one sold.

217 a, b
Movado "Kingmatic" wrist watch in an 18 ct water-resistant screwed gold case, made in 1960. The dial with centre seconds, signed "Chronomètre Movado Kingmatic", Reference No. 5111/12. The lever escapement 11¾''', Calibre 536 movement, No. 3002, with direct centre seconds, automatic winding from a central rotor, 28 jewels, precision adjusted.

* This information was kindly supplied by Herr H. Neumüllers.

Nathan George Horwitt

When the baby Nathan was born in 1898 near the White Russian town of Minsk, nobody could have foreseen that he would become one of America's foremost designers and that he would develop a revolutionary new watch dial. Three years after he was born, his Jewish parents found, as did so many others, that there was no future for them and their children in their own country, so they emigrated to the USA. Nathan George Horwitt grew up in the Bronx borough of New York, and attended high school, the City College and university in New York. He took part in the First World War as an American soldier and received his artistic training at the "Art Students League" in New York. After his first job as an advertising text writer and advertising manager in the 1920's, he became self-employed with the firm "Design Engineers Inc." after 1930. The industrial design firm designed digital clocks, radios, lamps, furniture, refrigerators etc. It was there that he designed a chair ("Beta Chair"), reminiscent of the tubular steel design by Marcel Breuer, an interchangeable all-glass picture frame and a digital desk clock named "Cyclox". Only the picture frame was a financial success and Horwitt had more than 2 million examples made in one of his own workshops.

The Bauhaus influence is clearly evident in Horwitt's work. The Bauhaus Movement, dissolved after Hitler came to power, exerted a great influence in the USA after 1933, when many of the leading exponents emigrated there. Men such as Walter Gropius, Ludwig Mies van der Rohe, Marcel Breuer and Laszlo Moholy-Nagy were responsible for the complete dominance of the style in architecture, industrial and furniture design. Horwitt also saw himself as a successor to the Bauhaus as he formulated his concept of design: "Functional, basic, not particularly respectful towards other styles or decorative forms, that are anyway only excuses for artistic sterility" or "Without use there is no beauty – without beauty – what is the use?" Apart from this Horwitt was a man of many talents and professions: writer, photographer, inventor, politi-

cian and farmer (in the last 40 years of his long life he farmed his own 400-acre farm in Lenox, Massachusetts) and he designed not only industrial products according to the functional and basic concepts but also books, advertising and interior spaces.

In 1947 Horwitt had the idea for a round watch dial devoid of any divisions or numerals, in other words completely plain except for a single dot at 12, and the time indicated by simple stick hands. A design which has its origins in his digital clock "Cyclox", patented in 1939, where his aspirations to free the product from unnecessary decoration and confine it to the most stark necessities, embodying the Bauhaus philosophy, are clearly definable. In 1955 Greta Daniel of the Museum of Modern Art made the following interesting comment to Horwitt:

"... I found that the continuously changing relationship of the moving hands to the 12 o'clock dot created attractive geometric patterns which were fascinating to observe. I realize that this esthetic pleasure can only be derived from a carefully balanced relationship of pure geometric elements...".

Horwitt attempted unsuccessfully to arouse the interest of Vacheron & Constantin to produce a wrist watch with this dial during a visit to Geneva in 1947. He applied for a US patent for this dial design in 1956 and after an initial rejection was granted Patent No. 183 488 on 9th September 1958. There was doubt expressed at the time that an (almost) entire emptiness, an (almost) total absence of design could be an important, patentable design medium. However, Horwitt was convinced of the value of his idea and was able to call in good references. The well-known photographer and director of the photographic department of the Museum of Modern Art, Edward Steichen, proclaimed that Horwitt's dial design was the only truly original and beautiful one for such an object. Douglas Haskell, the editor of the architectural magazine "Forum", wrote that exactly the reduction to a circle and a simple dot was revolutionary, truly new, the prin-

218
The dial of Horwitt's patented digital desk clock, "Cyclox", a forerunner in the development of The Museum Watch dial.

219
Portrait of Nathan G. Horwitt
with a prototype of the
Museum Watch, taken in
about 1960.

220 a, b
The second of three prototypes with Horwitt's patented dial, now on display in the Museum of Art in Brooklyn, New York.

ciple of "powerful emptiness" (whereby he borrowed a quotation by the Chinese philosopher Lao-tse). Thus the dot achieved decisive significance; without it the whole thing was no design at all.

Then Horwitt had three prototypes made through the New York representatives of Vacheron & Constantin and LeCoultre. They were simple, series made wrist watches in very plain white gold cases with LeCoultre movements, fitted with his patented dial; enamelled black with either a silver or gold dot. One prototype was accepted into the permanent collection of the New York Museum of Modern Art in 1960. Horwitt later presented the second prototype to the Brooklyn Museum of Art in New York and kept the third for himself. This second prototype has been on display in Brooklyn Museum since 1985, the Museum of Modern Art having taken its specimen out of the exhibition.

Between 1958 and 1960 Horwitt used the third prototype to try to find a watch manufacturer, either in America or in Switzerland, to produce wrist watches with this dial.

It is quite possible that already, some years earlier, Horwitt had had some prototypes made

by a New York watchmaker using Longines watches, which he had purchased because they had the very plain cases that he desired. Horwitt himself is not quite clear about this in his autobiography published later. Apart from this, the already quoted comments of Greta Daniel, dating from 1955, state that Horwitt showed her a wrist watch (or watches) including a lady's wrist watch. It seems hardly likely that she would have made her interesting comments without having seen an actual watch.

Horwitt's endeavours to utilise his design in a practical form advanced slowly owing to the moderate interest shown by the watch manufacturers. Many of them rejected the design, finding it too futuristic and risky. However, Horwitt was a tenacious man and he finally reached an agreement with Movado after 1960. Movado started to equip wrist watches from the standard programme with Horwitt's dial and sell them through a few New York shops. This beginning was very cautious at first, much too cautious for Horwitt who complained, "Movado at that time wouldn't spend a nickel on advertising." This changed a few years later when Movado began to advertise and in 1962 the dial design was registered as a trademark.

The term "Museum Watch" stems from the time when Horwitt's prototype was on display in the permanent exhibition of the Museum of Modern Art. However, this prototype was not a Movado wrist watch. The manufacturer was of little importance in this instance; since it was the dial alone that mattered. It was the dial's high aesthetic quality that made it one of the most important examples of the further development of the Bauhaus style in the USA, and the reason that it was included in the Museum of Modern Art exhibition. Movado took advantage of the fact that Horwitt's dial design was (and still is) part of the Museum of Modern Art permanent collection and used it to christen the watch and further the advertising of the product.

The integration of Horwitt's dial without numerals into the Movado standard programme was gradual. The first specimens were, as we have said, taken by the New York branch, one of the most important and managed by a Ditesheim. They were based on watches from the standard programme and only sold in the USA.

This situation did not cause any difficulties since complete watches were subject to a 35% customs import duty on entering the USA, and for this reason just Swiss movements with dials were frequently imported. The dials were often only temporary as the Swiss did not want to export dial-less movements to protect their own dial manufacturing industry. Therefore, most movements arrived with inexpensive provisional dials, just adequate enough to satisfy the regulations, which were replaced with the final dials in the USA. This was the general practice for simple wrist watches, not only at Movado, but throughout the Swiss watch industry. On the other hand, complicated watches with chronograph or calendar functions were imported complete. Since the cases and dials were generally produced in the USA it was possible to launch a model such as the Museum Watch from time to time, without it being part of the official company programme developed at the headquarters in La Chaux-de-Fonds. The headquarters clearly had initial reservations about this dial, which was so new and revolutionary. The large volume of production was seen as something of a gamble.

When it was accepted into the Swiss production programme in 1965, the first series was equipped with a very flat manually-wound movement, Calibre 245/246, which fitted in very well with Horwitt's concept of a flat wrist watch with similarly flat dial and crystal. This initial model was followed by many variations. The wrist watch with the almost completely empty dial and the dot at 12 became a commercial success. Horwitt's much-copied dial design contributed – and still contributes today – considerably to Movado's image. The familiarity and popularity of this dial is, as ever, undisputed; and one never ceases to be amazed at just how exactly one can tell the time from it.

221
1994 advertisement for the Museum Watch.

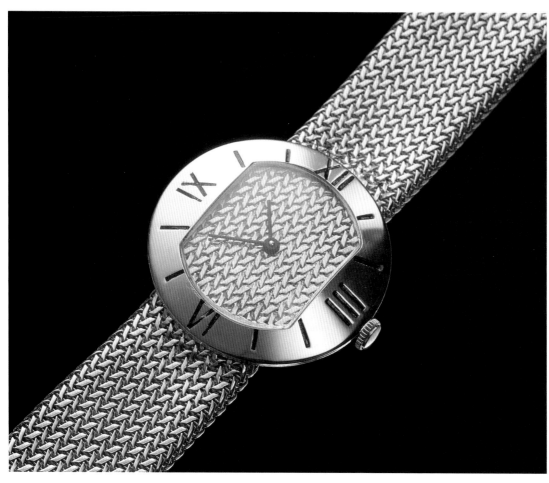

222
Dress wrist watch in an 18 ct gold case with gold link bracelet giving the optical impression that the latter continues across the dial, made in 1960. The bracelet and dial with a braided decoration, unsigned, no seconds. The numerals are on the gold bezel. The case is inscribed "UTI Paris – Movado – Spitzer & Fuhrmann", No. 40383. Very Flat 10"', Calibre 205 manually-wound movement with lever escapement, 17 jewels.

223
Extravagant gold horizontal oval wrist watch, made in 1965. The champagne-coloured dial signed "Movado", numerals on the bezel, Reference No. 255 33. The movement is an 8¾"', Calibre 246 of only 2.45 mm casing height (Universal Calibre 820) with lever escapement and somewhat increased vibrations (21 600 vibs./hour), 17 jewels.

224
Movado 18 ct gold-cased
wrist watch, made post
1966. The black dial with
gilded cartouches for the
numerals, no seconds, signed
"Movado", Reference No.
246 214 092. Manually-
wound 8¾''', Calibre 246
movement (Universal Calibre
820) with lever escapement,
17 jewels.

225
Movado 14 ct gold-cased
wrist watch, made in 1965.
Silvered dial, signed "Tiffany
& Co" below the Movado
trade mark, Reference No.
45333 214579. Nickel-
plated 10''', Calibre 205
manually-wound movement with lever escapement, No. 5729, 17 jewels, 2 adjustments.

226
Water-resistant steel-cased Movado
wrist watch, made in 1962. Tri-
coloured silvered dial with an inner 12-
hour chapter ring and an outer 24- hour
one, centre seconds, signed "Movado",
Reference No. 13089. Nickel- plated
11½''', Calibre 346 movement with lever
escapement, uncut monometallic balance wheel, flat
balance spring, 17 jewels, 2 adjustments.

**The path to
"El Primero"**

At the beginning of the 1960's it was once again time for a change of generation within the company management. The company had 325 employees by then; 50 more than in 1944. Georges Ditesheim (1893 – 1967), son of Isaac and President of the Board of Directors, as well as Roger Ditesheim, retired in 1959. At the beginning of the decade, the Technical Directors were Pierre (1896 – 1982), the son of the company founder, Achille, and Bernard (born 1929), the son of Georges. The Commercial Directors were Armand, Lucien, Edouard, Alain and Bertin Ditesheim. In 1962 Pierre retired after 42 years of service in the firm, but he continued to concern himself with technical matters even in retirement. In his latter years he was particularly involved in the introduction of electronics to watchmaking. His place on the Board of Directors was taken by Armand Ditesheim, who came from the commercial side. In 1967 Lucien Ditesheim, the chief financial director, retired. How strongly the Ditesheim family directed the actual business of the company may be judged by taking a look at the composition of the Board of Directors in 1965: 9 Ditesheims and 3 non-Ditesheims (Raymond Polo, Rémy Krähenbuhl and Henri Wagner).

This time of generation change at Movado coincided with the start of changes in the Swiss watch industry itself, with serious consequences for them both. For the 1960's saw the development of electric and electronic drives for wrist watches, which also had new frequency control systems. First came the electro-dynamic balance in 1960. This was followed a year later by the tuning fork as the frequency controlling element. However, both systems made only a relatively brief appearance on the market after the breakthrough of quartz frequency control for wrist watches came in 1970. Due to low service requirements, convenience, high accuracy and eventually very low price, these wrist watches were superior to their mechanical counterparts. This superiority grew with the ever-increasing perfection of the new technology. The effect was

that manufacturers which had concentrated entirely on the production of mechanical watches – almost the entire Swiss industry – gradually experienced economic difficulties towards the close of the 1960's caused by the increasing competition. Movado was not spared from this development. In fact, the Swiss watch industry was particularly hard hit for the following reasons: firstly, the watch industry was of major importance in Switzerland; secondly, Swiss manufacturers dominated the world market with their mechanical watches; and thirdly, these new technological developments were taking place elsewhere, notably in the Far East and the USA. The threat of missing the boat loomed large for the Swiss. The fact that the 1960's had been one of the most successful periods in the history of the mechanical watch also played a rôle and the Swiss tended to bask too long in their success.

To ward off the competition from the new impulses, immediate plans were implemented to make the mechanical wrist watch more competitive with ease of service, better value for money and improved accuracy. The result of these developments was the so-called fast-beating movement with a lever escapement and a balance frequency of 36 000 vibs./hour, which gave a better and more stable rate due to the higher frequency. Almost all of the major watch manufacturers had such watches in their programme by the end of the 1960's. The Movado version was the "Kingmatic HS 360", with movements from the 404 – 409 Calibre group. A water-resistant steel case, automatic winding and high precision fast-beating movements made these extremely good wrist watches, with rates comparable to tuning fork watches, and only exceeded by watches with quartz movements. They were capable of running for many years without service and were fully autonomous whereas the first series of electronic watches required a battery change every few months. The mechanical watches gave better value for money in the first instance and cost about the same in the long term. It would seem that they really were com-

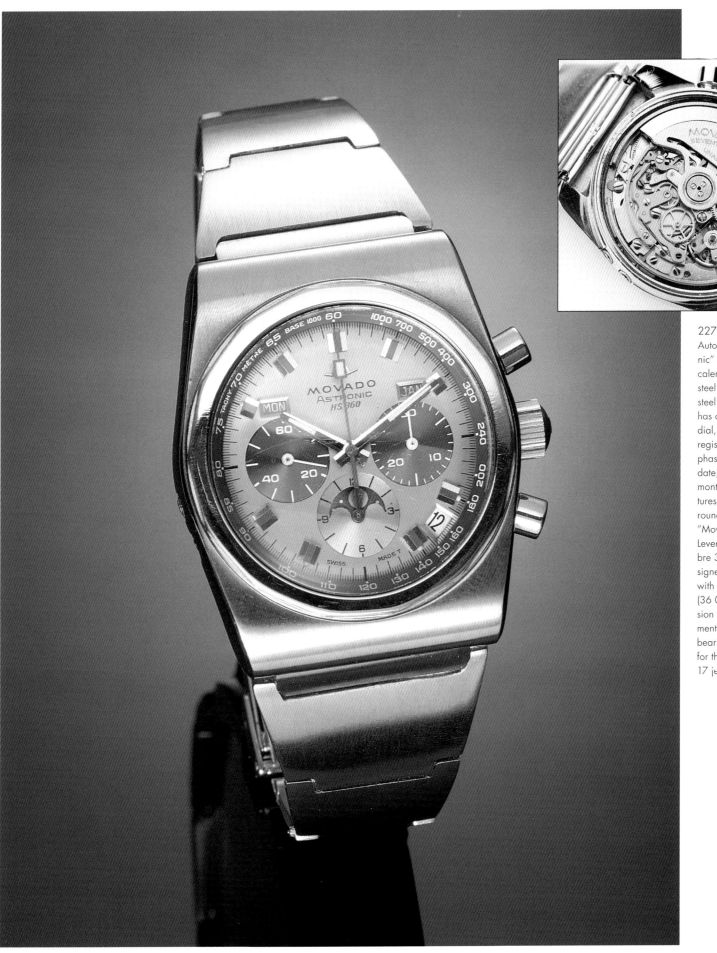

227
Automatically wound "Astronic" wrist chronograph with
calendar, made in 1970. The
steel case is with its original
steel link bracelet. The dial
has a small running seconds
dial, 30-minute and 12-hour
registers, the latter with
phases of the moon display,
date, day of the week and
month displayed in aper-
tures, tachometer scale surrounding the dial, signed
"Movado Astronic HS 360".
Lever escapement 13''', Calibre 3019 PHF movement,
signed "Movado Factories",
with fast vibrating balance
(36 000 vibs./ hour), precision regulator index adjustment via an eccentric, ball
bearing-mounted central rotor
for the automatic winding,
17 jewels.

petitive with the electronic wrist watches. However, the trend could not be reversed as, thanks to clever advertising, the electronic watch had the image of "progress", against which reason cannot win, as is well known.

In another attempt to master the ever increasing economic difficulties, Movado tried cooperation – restricted as far as time and projects were concerned – with one or several other companies, since joint research and development could save a lot of money and joint use of either movements or components would create additional savings. Movado already began to work with Universal in the early 1960's, their aim be-

228
View of the Calibre 3019 PHC movement used in the "El Primero" and "Datron".

ing to develop and produce a joint Calibre. Movado abandoned the manually-wound Calibre 365 and took over the slightly smaller Universal Calibre 1105 as well as the 1‴ smaller 1005, which became Movado Calibres 345 and 335 respectively. Universal, on the other hand, took over Movado rough movements of Calibres 5, 6, 45, 54, 56 and 395 E. The latter was given the confusing designation "Microtor" at Universal although it had a normal central rotor. A little later, in 1966, Movado took advantage of the flatter automatic movement with an offset micro-rotor developed by Universal as Calibre 66. This 2.5 mm high micro-rotor movement was known as Calibre 230 at Movado.

Shortly after this Movado changed partners. A joint venture with Zenith had the ambitious target of producing the first series-produced automatically-wound chronograph which would also have a date indication. This develop-

ment, the final peak of the mechanical wrist watch before the onset of the quartz revolution, was too difficult and complex for the resources of a single company and Movado's rich experience of complicated wrist watches, gained over a long period of time, made it an ideal partner for such an enterprise. Somehow the idea of an automatic with chronograph must have been in the air since a whole row of other firms were working to realise it. The main competition came from the amalgamation of the chronograph specialists Breitling and Heuer-Léonidas with Hamilton-Büren and Dubois Depraz, they had started development of their own project in 1965. Both groups achieved their aims independently, at about the same time, and exhibited their new developments at the Basel Fair in 1969. However, the Movado/Zenith model was accepted as the first on the market whence the name "El Primero" is derived.

The automatic winding of the Movado/Zenith watch has a centrally positioned rotor. The date indication is in an aperture positioned between 4 and 5. The chronograph is equipped with 30-minute and 12-hour registers and as well as the normal hour and minute hands, there is a small seconds hand at 9 and a tachometer scale around the edge of the dial. The 13‴ movement of Calibre 3019 PHC with an overall height of 6.5 mm, which is particularly flat in view of the various complications, was a completely new development. The balance beats 36 000 vibs./hour and there are 17 or 31 functional jewels and a rotor running on a ball bearing. These specifications made it the most modern watch of its type on the market. "El Primero" earned an excellent reputation due to its high quality.

There are several case and dial variants of the El Primero signed with either Movado or Zenith. There was an additional version, technically almost identical, but with the date aperture at 12. This was the only difference between the El Primero and the "Datron HS 360" which was signed exclusively Movado. Yet another variant was known as Calibre 3019 PHF with a full calendar. In addition to the date aperture between 4 and 5 there were further apertures for the day of the week and the month with the phases of the moon shown within the 12-hour counter chapter ring. This model, designated the "Astronic HS 360", was often fitted with massive and clumsy steel or gold cases that were typical of that period and frequently twice as thick as was required for the movement.

229
Automatically-wound Movado "Datron" wrist chronograph with date indication in a 14 ct gold case, made in 1970. Gilded dial with small black dials for small seconds, 30-minute and 12-hour registers, external tachometer scale, date indication in an aperture at 12, signed "Movado Datron HS 360", Reference No. 434 225 502. Lever escapement 13‴, Calibre 3019 PHC movement, signed "Movado Factories", with fast vibrating balance (36 000 vibs./hour), precision regulator index adjustment via an eccentric, ball bearing-mounted central rotor for the automatic winding, 17 jewels.

1969, the year that had begun so successfully with the launch of the joint venture the "El Primero", marked the end of Movado as a family business run by the Ditesheims for the past 88 years. The "marriage of convenience" with Zenith was consolidated in September 1969 by a permanent union of Movado-Zenith-Mondia (MZM) as a holding company managed by Zenith as the majority shareholders. Mondia was a watch assembly company in Sierre, Canton Valais. By 1971 Movado had gradually moved all activities to Le Locle, where Zenith had been since the company's founding in 1865. A new factory building was erected there; and while the current products, for example the mechanical Calibre 3019 PHC were still manufactured, quartz watches soon dominated the production as MZM strove to keep up with the changed market. Movado also adopted the Zenith numbering and reference system in 1971.

Looking at the sales catalogue from this time, one sees that many of the watches bear the company name "Movado", others "Zenith-Movado" and some occasionally also "Movado – a Zenith Company". One also receives the impression that the Museum Watch with Horwitt's dial design remained one of the best model lines. There were numerous variations of it with a Movado mechanical movement or Zenith Calibre 2320 as well as quartz movements by Movado or Zenith, sometimes signed with both company names. The gold dot at 12 had developed into a date aperture or carried a diamond solitaire (the most expensive variant with an 18 ct white gold case and diamond-set hands). Cases were rectangular or oval, the dot silver coloured with steel or white gold, and finally, the dot dial design appeared not only in wrist watches, but also in a rectangular desk clock with an 8-day movement.

Other renowned Movado models such as the Queenmatic, Tempomatic, Kingmatic and Kingmatic HS 360 were also continued with classical and contemporary modern cases and dials e.g. a Kingmatic version called "Video", which featured a horizontal oval case and dial with the proportions of a television screen. There were also the numerous lady's wrist watches, with simple or elaborate decoration, that were traditional Movado lines. The leading model was the "Datron" chronograph with automatic winding, available with a number of different dials such as the "Chronodiver" with a rotatable bezel for

230
Automatically-wound "Datron" chronograph with date, diver's version, made in 1970. The steel case with protected winding crown and its original steel link bracelet. Blue dial with silvered dials for small seconds, 30-minute and 12-hour registers, rotat- able bezel, Reference No. A 34705 501. Automatically-wound 13''', Calibre 3019 PHC movement with lever escapement.

231
Four Movado-Zenith quartz wrist watches with various versions of the Museum Dial, post 1970.

232
Extract from a 1970's Movado catalogue showing watches with Museum Dials of various colours.

21.027.1.35.2

7.2.35.2

20.027.2.35.2

c106   c107   c108   c109

233
Some "Video" line models from the 1970's.

42 **MOVADO** Video HS 360 Day and Date automatic Sub-Sea

divers, the "Datachron" without any noticeable difference from the Datron, and the "Astronic HS 360" with full calendar. Despite Zenith's economic dominance in this holding, it would appear that Movado's watchmaking individuality was largely maintained.

This programme also appeared to change very little when the holding was bought by the North American "Zenith Radio Corporation of Chicago" in June 1972, and was active as a subsidiary company under the name "Zenith Time SA". The American radio and television manufacturer Zenith Radio had no connections with the Swiss watch manufacturer Zenith prior to 1972, other than paying a licensing fee for use of the name Zenith. However, times became harder for Movado under the Zenith Radio Corp., which was foreign to the branch: it was the era of cheap quartz watches manufactured in huge quantities. The international production figures for mechanical and quartz watches combined rose from 218 million in 1975 to 700 million in 1989. The situation did not begin to change for Movado until Zenith and Movado were once again taken over by a Swiss concern in June 1978. This was Dixi SA under Paul Castella, together with the Englishman Michael J. Pannett as the majority shareholder.

The Dixi company was founded at the end of the 19th century in Le Locle by the watch manufacturer Le Phare, for the production of machine tools. However, Dixi soon developed into the most important branch within the company and gained increasing independence. Through its purchase of the watch brand name "Paul Buhré" in 1963, followed by "Luxor" and "Zenith-Movado" in 1978 and "Zodiac" in 1979, Dixi itself became a major watch manufacturer. The company, now called "Zenith Movado Le Locle SA", sold its watches to the North and South American markets under the brand name Movado (Movado Time Corp.) and in Europe under the name of Zenith, which was better known there. The most important dealers in the USA were Tourneau in New York, Marshall Field's in Chicago, Neiman Marcus in Dallas and Polacheck's in Los Angeles.

By 1980 the only movements still manufactured for wrist watches were quartz ones. There were sufficient stocks on hand in the stores for the few models still available with mechanical movement for nostalgic customers, e.g. the wide-ranging group of Museum Watch models. One of the typical models from these years was the flat quartz watch "Centenaire" to mark the centenary of the company. This event promoted an in-depth article in the February 1981 issue of the American "Jewelers' Circular Keystone". The story profiled the then manager of Movado-USA, Gerry Hansen, assuring readers that Movado was on the way up again now that it was back under Swiss management. The sales figures had increased noticeably and the advertising budget had been increased by 200% within two years, according to the article.

By 1981, ultra-flat quartz watches were being made: the "Delirium" – only 1.98 mm thick and ever-new variants of the Museum Watch e.g. the "Museum Imperiale" with a steel case and an integrated steel bracelet decorated with gold dots. For this and other models a black cat was used in the advertising, and black was generally used a lot. Yet the Museum Watch dial was being produced not only in black but also in red, blue, and brown, since it had been observed that women in particular wished to have the watch in different colours. Horwitt himself took part in this development.

234
Movement of the white gold-cased lady's wrist watch shown in Illus. 236. Tonneau-shaped form movement, Calibre 1110, with lever escapement and a large uncut monometallic balance wheel, signed "Movado F'ries", 17 jewels.

235
Movement of the gentleman's wrist watch shown in Illus. 236. Manually-wound 11½''', Calibre 2552 C movement with lever escapement, signed "Movado Factories", precision regulator index with an eccentric, 17 jewels.

236
Two classical round Movado wrist watches with Museum Dials and mechanical movements, made in 1975. Left, a lady's wrist watch cased in 14 ct white gold, right, a gentleman's wrist watch cased in 14 ct yellow gold, with the date shown in the dot at 12 and a black onyx set in the winding crown.

237
Movado quartz wrist watch "Centenaire", created to celebrate the centenary of the company in 1981.

238
Two oval Movado gold-cased wrist watches with Museum Dials, made in 1975. Left, the gentleman's wrist watch with a quartz movement and matching oval dot on the dial. Right, the lady's wrist watch with a diamond set in the dot at 12, signed "Movado Zenith". Mechanical 7¾''', Calibre 1740 movement with lever escapement, signed "Zenith".

239
1981 Movado advertisement for the quartz wrist watch model "Museum Imperiale".

Acquisition and a new start

240
Gedalio (left) and
Efraim Grinberg.

It was a revolution that sent Gedalio Grinberg to the USA. In 1959, after a three-year struggle, Fidel Castro and his rebel army overthrew the government of Cuban President Batista, and took control of the country. Grinberg, who was born in Cuba in 1931 and grew up in a family active in the jewellery trade, realised that he had no future in his homeland. For in Cuba – as in all other so-called "people's republics" – only Government supporters, in this instance Castro followers, could get ahead and Grinberg refused to co-operate with them. Aware of the difficulties he would have to face if he remained in Cuba, and shortly after having been picked up by Castro's security agents and temporarily held in custody, he headed for Miami.

At the time when their son was born in Cuba in 1931, Gedalio Grinberg's parents already had a life behind them that we would call a modern odyssey today. His father had fled from post-revolutionary Russia to Palestine, where he met and married a young refugee from Poland. However, the hard and dangerous life there, as well as malaria, soon caused the attractions of the promised land to fade and the young couple took a boat to Cuba in 1930. This was intended to be a temporary resting place as they really wanted to go on to the USA. But, by the time their US visas were granted, the Grinbergs had already established ties in the small Cuban village of Quivican, where they had stopped en route to the USA, and so they decided to stay there. Grinberg senior opened a business for cloth, jewellery and watches. This rather involved emigration is somewhat reminiscent of the very much shorter and more comfortable one undertaken by the Ditesheim family 50 years earlier. The Ditesheims had founded Movado and steered it to greatness; the Grinbergs were to rescue Movado from the anonymity of frequently changing partnerships and restore its status as an independent brand.

After 1944 the family moved to Havana, where Grinberg senior opened a jewellery store. After passing his baccalaureate examinations in

1949, Grinberg junior worked in his father's business for a while and studied economic science at Havana University. His interest in the watch trade had long since been awakened. In 1954 he became the partner of one of his father's suppliers, Fabian Weiss, the representative for Omega and later Piaget, in Cuba. The business was very successful; during the 1940's and 1950's Cuba exercised a great attraction on American tourists who willingly spent their dollars there. During this period, Ernest Hemingway had a house in Cuba and set off from there on one of his favourite occupations, deep sea fishing. Cuba and fishing inspired him to write one of his best known novels, "The Old Man and the Sea", which was published in 1953.

Then came Castro's Revolution. In 1960 Grinberg, his wife, two small children and his partner Fabian Weiss moved to Miami. Very soon after, through the good offices of Camille Pilet, who was then the international sales manager of Piaget, that luxury Swiss watch firm offered them the exclusive sales rights for their products in the USA and Puerto Rico. Taking up the offer, Grinberg together with Weiss and Weiss' son José opened the New York branch office of Piaget in 1961. Initially, they faced a number of difficulties. All three co-partners were newcomers to the USA, and Manhattan was a challenging city where business competition flourished. Moreover, the Piaget brand was virtually unknown in the USA.

The situation improved significantly when Grinberg was able to convince leading retailers, such as Neiman Marcus, Van Cleef & Arpels, Tiffany & Co. and others, to carry the Piaget product line, and, as was the case with Van Cleef & Arpels, to establish a fine watch department.

In 1965 Grinberg purchased the Weiss family's interest in their business, and founded the North American Watch Corporation with the vision of importing and distributing other fine watch brands in the future. The acquisition of trade rights for further Swiss companies soon followed. First of all Corum, a very young company founded only in 1955 in La Chaux-de-Fonds, which specialised in the manufacture of high-grade watches. Already in 1969 when Movado lost its autonomy and merged with Zenith, Grinberg tried to acquire the American branch of Movado. This company had a well-organised and widely networked sales organisation in the USA and, from time to time, even followed its own inclinations as is shown by the Museum Watch. At that

time, and until 1972, the American branch of Movado was headed by Gerard Ditesheim, Isaac the engraver's grandson, and Grinberg was not able to achieve his goal. So in 1970, he first purchased the watch brand Concord, and this time it was the entire company, not just the regionally restricted sales rights as hitherto.

Concord, founded in Bienne in 1908, was quite unknown as an independent brand in the USA, for at that time it principally made watches for the private label sector; which means that individual dealers placed their own names on the dial and the name of the manufacturer was only to be seen on the movements. In the 1970's Concord made good sales within NAWC, with some models of its own, in particular the quartz wrist watch "Delirium", launched in 1979 as the flattest wrist watch in the world. However, Concord's sales figures significantly decreased in the early 1980's, during a worldwide recessionary phase which also affected Zenith Movado Le Locle SA, part of the Dixi concern. During the summer of 1982, Grinberg along with his son Efraim, who had joined the company in 1980, started negotiations with Dixi's Director Castella to purchase Movado. In 1983, contrary to the advice of all the financial experts, the Grinbergs were successful in taking over Movado.

Gedalio Grinberg had always been fascinated by Movado, by the timeless elegance of their watches, and in particular by the Museum Watch, whose concept he admired and whose creator, Horwitt, he knew personally. Now, at last, he could realise what had been briefly and aptly described as "Making Movado" in a "Modern Jeweler" article of 1988. It took three years, 1984 being the most difficult and with the greatest losses incurred, to re-organise Movado, to give it new goals as well as an independent image and to bring it back onto the market. The first priority was to develop a new product line and then to create exciting advertising to promote it.

Advertising has a high priority at NAWC and is taken very seriously. This attitude is shown by the fact that the company has its own advertising department, in which concepts are developed and where as much advertising copy is produced as possible.

With a total advertising budget of around $25 million in 1987, $10 million was used for Museum Watch models alone. In 1994, NAWC ranked fourth in terms of advertising outlay among the American watch distributors (after Seiko, Timex and Citizen).

Grinberg was aware that with Movado, not only had he acquired a company with an unmistakable image and great technical know-how, but also one with an eventful history which had made a significant contribution to the history of watchmaking.

In order to build up the brand again it was of particular necessity to re-define the image of "Movado" and to identify and target potential buyers who differed noticeably from those of other brands. Broad-based marketing studies were carried out for this purpose, which defined the profile of this target group as follows: mostly young buyers (25 to 40 years old) with higher education, an interest in culture and the applied arts, and an awareness of their obligation towards society. The new Movado programme was accordingly established in the medium price range.

The latter, however, had got somewhat lost in the years before 1983, with the exception of the Museum Watch, which was well-known in the USA and whose idea actually originated in the USA and not in the Swiss headquarters. In order to take advantage of this situation, the Movado programme was largely based upon the Museum Dial: at first half of the production consisted of wrist watches with the Museum Dial. Even today, the greatest part of Movado's turnover comes from this watch type.

In addition to the classical Museum Watch, a great number of models with Horwitt's dial were developed: in the round version as a jewel-studded dress watch with gold or gilded expanding bracelet, in sporty versions with steel cases and bracelets, which the American tennis star Pete Sampras helps to advertise, and finally with a sapphire crystal and black satin finish case and bracelet. The Museum Dial is available not only in the traditional black, but in white, green, gold and grey, the latter combined with a waffle or mesh pattern. The different combinations of watches with the Museum Dial are almost endless, and new variations are constantly being designed. Nearly all of them have a quartz movement; one exception will be described later.

Since the second half of the 1980's, Movado has distinguished itself as the manufacturer of special products (see the watches of the Artists' Series) and also as the sponsor of cultural events. The company is a major benefactor of American Ballet Theatre (ABT) in New York. And at the beginning of the 1990's, Movado supported a multi-part television programme made by the public television stations, entitled "Art of

the Western World". The programme is still used as a teaching aid in schools of further education. As previously mentioned, thanks to lavish and clever advertising (in all the appropriate media), Movado wrist watches with the Museum Dial are so well-known that most people think Movado only makes this watch and has never made any others. The company plans to combat this view in future with an intensified extension of the programme towards non-Museum watches (without neglecting the Museum Watch, of course). This erroneous impression is also refuted by the many other models and lines of models, which will be described in more detail below.

This extension of the company's programme, as well as the opening up of other markets, e.g. in Europe and in the Far East, is the aim of Gedalio Grinberg and his son Efraim, who has been President of NAWC since 1991. The turnover figures alone show to what great success the Grinbergs have brought Movado: in 1983, the year of their acquisition of the company, the turnover amounted to $4 million; in 1987 it was already $50 million; and in 1994; almost $100 million. How modest, in contrast, the turnover of 1881 seems, with its 40,000 Swiss francs!

241
Pete Sampras

Organisation

Here an overview of Movado's organisation today. With its purchase by Grinberg in 1983, the company management definitively moved to the USA and, as mentioned, became part of the North American Watch Corporation, whose headquarters were initially situated on Manhattan's Fifth Avenue. In 1987 Movado was separated and established not far away in Lyndhurst, New Jersey, on the other side of the Hudson River. Today the entire NAWC organisation is together in Lyndhurst. There is still, however, a "Movado Design Store" at 630 Fifth Avenue in Manhattan. On the ground floor you can admire the whole range of Movado products. On the upper floor, part of the extensive collection of historic Movado watches is displayed, making it the only dedicated watch museum in New York. In Lyndhurst, apart from the company management, there are the advertising department, the central sales management, a development and design department, dispatch, the legal service as well as the central repairs department.

In October 1993, NAWC made a successful stock offering and became a public company. Based on this strategically important move,

View of a display case on the upper floor of the "Movado Design Shop" in New York.

Movado was subsequently able to increase its equity to over $100 million dollars.*

The production of Movado watches takes place in Bienne, Switzerland. There, under the management of Kurt Burki, about 120 employees are engaged in producing watches of the Movado and also Concord brands, assembling them mainly from bought-in parts. Today these watches mostly have quartz movements. The cases and metal bracelets are largely produced there, too, since the case- and bracelet-maker Grandjean is also part of the Swiss branch.

In addition to production, quality control (in which all bought-in watch parts as well as the finished watches are tested extensively according to sophisticated processes), the development and design department, the sales management (via agencies in more than 60 countries), the Swiss headquarters at Rue Centrale 63 in Bienne also houses an after sales service facility. This is where spare parts are kept at the ready and repairs carried out, and where training courses for managers, salesmen and watchmakers are held, during which technical knowhow, service and sales techniques are taught. It is also from here that course participants are visited and advised in their local places of work.

And it is in Bienne – in consultation with the company management in Lyndhurst – that most new models, such as the "Collection 1881", which was inspired by classical models and is to be described in detail later, and many variants of the Museum Watch are developed. It is also where the prototypes and trial runs of new watch models are made and then extensively tested to ensure that the usual high quality is upheld before they go into production.

* In April 1996, NAWC changed its corporate name to Movado Group, Inc., to better reflect the international scope of its business. A subsidiary Company retained the name North American Watch, and cotinues to distribute Movado watches, as well as the corporation's other four brands, throughout the US, Canada and Caribbean.

242
Movado sponsors the "ABT" (American Ballet Theatre).

The new Movado models

In addition to the models with the Museum Dial, a new range of models was developed after 1983. The most extravagant and interesting of these is the series of watches designed by artists and named the Movado Artists' Collection. The concept was to invite either a well-known, established artist or a promising, young newcomer to design a watch quite freely, according to his own personal concept of time. Movado would produce a limited number of each such watch. Since 1988, with the exception of 1992, there has been an artist's watch each year – a total of six unique designs. So far, the modest total production of these watches (2,479 pieces in all) stands in a strange contrast to their great significance and degree of fame – and to the amount of space they will take up in our description.

The idea for this most unusual range of models came from Gedalio Grinberg, himself an avid collector of modern art; in his collection there are, among others, works by Paul Jenkins, Ruth Nevelson, Wilfredo Lam, Yaacov Agam and – Andy Warhol. Warhol, a personal friend of Grinberg's, was the first artist to create such a watch for Movado, the Andy Warhol Times/5, which appeared shortly after his death in a production of 250 pieces. It was described in "Alte Uhren" by Gisbert L. Brunner as "unusual and striking". About six weeks before his death, Warhol had shown Gedalio Grinberg some snapshots of Manhattan that he planned to use for the watch, and his final choice of pictures was found in an envelope after his death. Warhol was a keen watch collector and frequent guest at all New York watch auctions; in his collection there were several vintage Movados, including 6 Ermetos.

Andy Warhol, the son of Czech immigrants, who was born in 1928 or 1930 and grew up in Pittsburgh, was the most famous and at the same time the most controversial of all pop artists. Warhol was not only a painter, but also a film maker, author, publisher and photographer. The idea of designing a watch, therefore, did not only appeal to him, it was also a natural choice.

Warhol had already thought of making a watch with several dials in 1984, and soon set to work on the logical next step of painting these. Given that he was a fanatical New Yorker as well as a great photographer, the very unusual result of his design process should not come as a surprise: five separate rectangular, black high-grade steel cases linked to one another by joints and forming a self-contained bracelet. And in each of these five cases there is an individual watch with its own movement, etc. In place of dials there are five different black and white photographs showing motifs from the cityscape of Manhattan. In the centre of the pictures two red hands seem to be floating in a more or less disconnected fashion; there are no numerals or other orientation marks on the photos. It is solely the arrangement of the hand-setting buttons that indicate where the twelve might be, although this is relatively unimportant, as anyone can place the division of hours on the wholly neutral dial as they wish. In their radical simplification Warhol's dials go much further than Horwitt's Museum Dials, which give a clear orientation with a dot instead of the twelve.

The "Andy Warhol Times/5" is a much sought after watch fetching a five-digit sum at auctions. One of them, for example, was knocked down at £ 14,950 at Christie's in London on 15th June 1994. And such a high price was probably paid for the work of pop art and certainly not because you get five watches for your money; five watches in one, from which not one can be removed without destroying the watch, making it useless as a complete watch. These five rectangular watches in a row thus form an indivisible whole, one single watch.

Andy Warhol Times/5

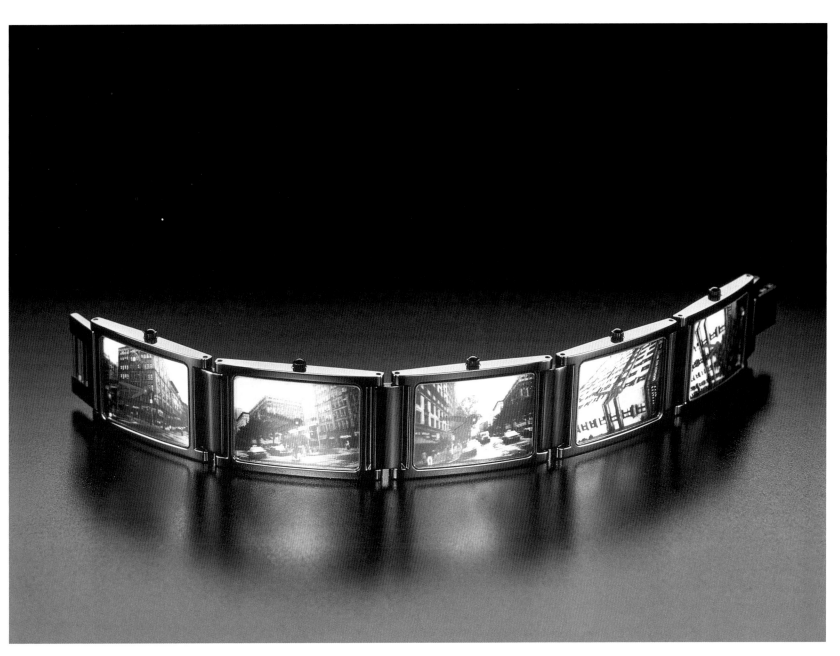

243
Movado Artists' Watches:
The Andy Warhol Times/5.

Yaacov Agam's "Rainbow" Collection

In 1989, Movado's second watch in the Artists' Collection, the four-part "Rainbow" Collection appeared. It was designed by Yaacov Agam, who was born in 1928 and lives in Paris.

This collection consists of four watches with a single design concept: a pocket watch (in a production of 250 pieces), a small clock (likewise 250 pieces), a wrist watch with a leather strap (150 pieces) and a wrist watch with a steel bracelet (100 pieces). Agam's concept for the dials consists of coloured rings following the sequence of the colours of the rainbow. These coloured rings are applied to two thin glass discs so that when they rotate a semi-circle and a closed circle are formed once an hour. Two rows of dots of differing length, also in the colours of the rainbow, serve as hands and indicate the time on an outer row of dots. The whole dial is a sophisticated interplay of constantly changing colours and geometrical patterns, vaguely reminiscent of Ernest Borel's Cocktail dial from the 1950's. Agam also designed two further collections at this time: the "Love Star" and the "Galaxy" Collections, which only differ from the "Rainbow" Collection with regard to the dial decoration.

Agam, a pupil of the Bauhaus master Johannes Itten in Zurich, may be counted alongside Calder, Tinguely and Vasarély amongst the most important representatives of op art and kinetic art, of which this watch collection is a typical example. And hardly anything is better suited to a kinetic work of art than the representation of constantly changing, elapsing time.

244
Movado Artists' Watches:
A desk clock from Yaacov Agam's "Love Star" collection.

245
Movado Artists' Watches:
A desk clock from Yaacov Agam's "Galaxy" collection.

246
Movado Artists' Watches:
Yaacov Agam's "Rainbow"
collection.

Arman's "The color of time"

Armand Pierre Fernandez, known by his artist's name of Arman, was the third artist – born like the others in 1928 – given the opportunity to design a watch for Movado. His design, a single and almost conventional wrist watch compared with its predecessors, was manufactured in 1990 in a production of 200 pieces.

For his concept, Arman used the tools of the painter: brush and paint. His wrist watch "The color of time", which has a manually wound movement, boasts a dial composed of rotating discs, on which two brushes of different length function as hands indicating the time on variously coloured rough brush strokes. The basic motif of the brush strokes is repeated on the strap and case of this watch.

Arman's symbolic use of his professional tools for the indication of time is typical of his surrealistic and neo-realistic style. He also assembles objects and "accumulations", for example from car parts or demolished pieces of furniture, giving them a new meaning.

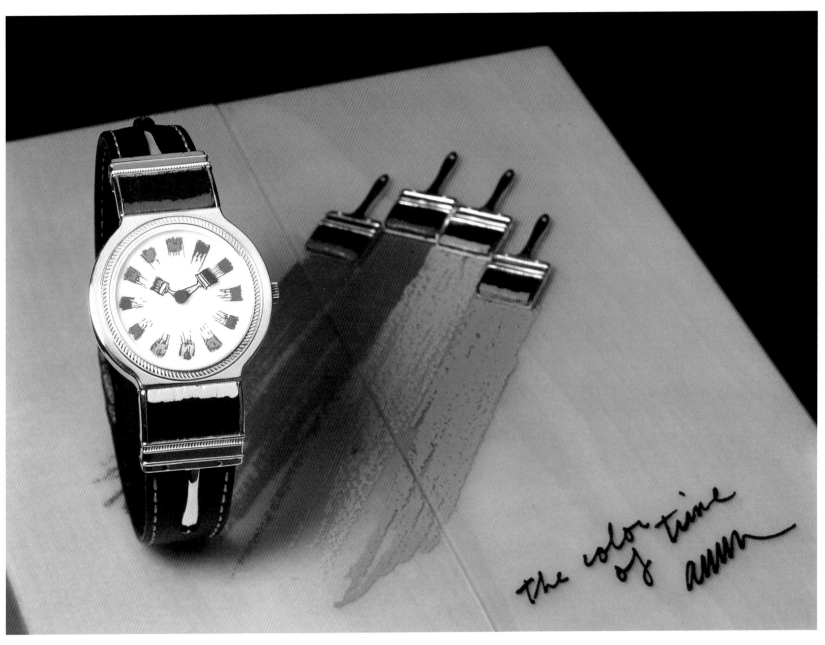

247
Movado Artists' Watches:
Arman's "The color of time".

James Rosenquist's "Elapse, Eclipse, Ellipse"

Rosenquist, born in North Dakota in 1933, began his career after studying at the College of Art in Minneapolis by painting billboards and petrol pumps. In this way he learned a lot about painting techniques and dividing up space. He is considered to be a representative of the new, abstract expressionism.

Rosenquist's wrist watch, which came out in 1991 and was limited to 180 pieces, consists again of several watches, but in a totally different way from Warhol's. The name "Elapse, Eclipse, Ellipse" refers to three watches running into one another – in such a way that the two outer and normal round-shaped ones symbolically squeeze the inner one into an irregular and flattened shape. However all three fully retain their individuality with quite differently painted dials. Rosenquist describes this dial painting as "floating pictures". It is a world-time watch, showing the time in three time zones. Rosenquist was thinking of local time in New York, Europe and Los Angeles, which are important for him. It is thus surely the most ambitious and most unusual design for a world-time watch ever made.

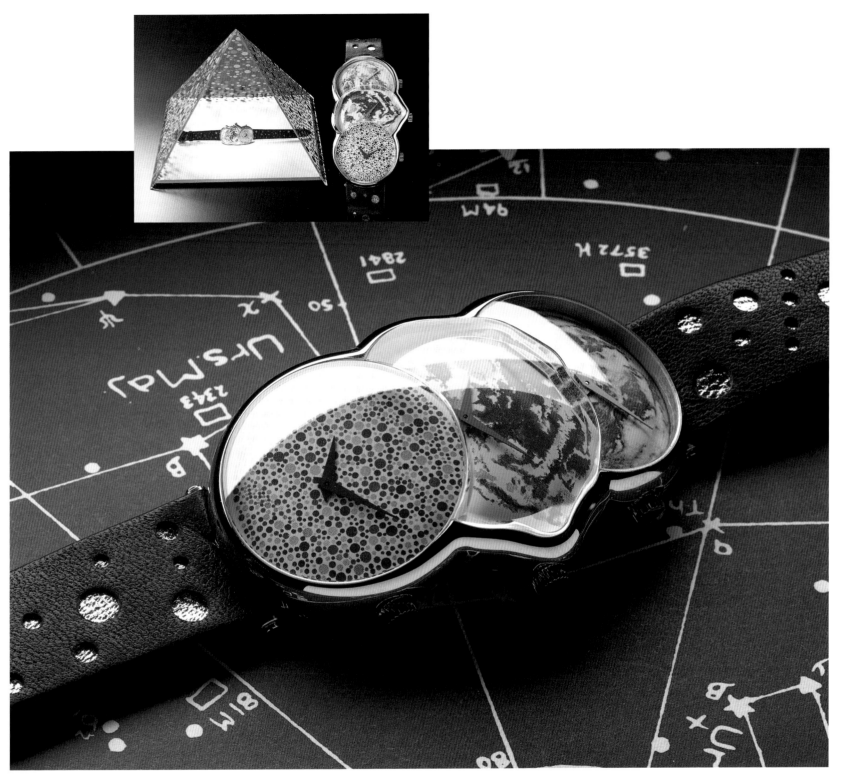

248 a, b
Movado Artists' Watches:
James Rosenquist's "Elapse,
Eclipse, Ellipse".

Max Bill's "Bill-time"

After a year's pause, the next Artist's Watch was the "Bill-time" designed by the Swiss artist Max Bill and issued in a production of 99 pieces.

Bill, born in Winterthur in 1908, is so far the oldest artist who has worked for Movado. The painter, architect, designer and teacher Max Bill studied at the Bauhaus in Dessau, has been working as a free-lance architect since 1929, and was one of the founders of the famous College of Design in Ulm, which saw itself as the successor of the Bauhaus in post-war Germany. He was the school's first principal from 1953 to 1957. In 1957, he left the school and designed several watch models for Junghans, the biggest watch company in Germany. There is also a further wrist watch design of Bill's for Junghans dating from the early 1960's. Thus Bill was no newcomer to the design of wrist watches when he received the commission from Movado.

The octagonal case and link bracelet of his watch are made of silver and, as in Arman's versions, they are finely graduated in the colours of the rainbow to achieve a harmonious unity between form and colours. Nevertheless the total effect remains cool due to the use of silver (Max Bill was formerly a silversmith) and cool colours. In the centre of the convex-ground sapphire crystal, there is a large silver dot from which simple baton hands radiate. The watch has a mechanical movement, which can be viewed through a glazed back.

The "Bill-time" is supplied in a double glass cylinder, also designed by Bill, combining to make such a beautiful work of art that – as Thomas Kapteina surmises in the magazine "Armbanduhr International" 3/1993 – "It is hardly likely that any of the owners will ever take the watch out and wear it on his wrist."

249
Movado Artists' Watches:
"Bill-time" by Max Bill.

Romero Britto's "The Children of the World"

The Artist's Watch designed by the Brazilian Britto in 1994 added a new dimension to this series: that of patronage, a foundation for charitable purposes. In this connection Gedalio Grinberg said: "Movado and its parent company North American Watch Corporation have both achieved much, and with success comes responsibility to provide equal opportunities for all people."

Movado will therefore donate the proceeds from the sales of this watch, limited to 1,000 pieces, to St Jude Children's Research Hospital in Memphis, Tennessee and to the worldwide children's charity, UNICEF.

Britto, born in 1963, lives in the USA today and is counted as a pop artist. This is also quite clearly shown by his wrist watch, as bright and colourful as a Swatch. It is presented in a decorative, transparent plastic statue, whose shape resembles the Chinese character "Chi", the symbol of human energy. This connection is typical for Britto, who also works actively on behalf of humanitarian and social causes.

250
Movado Artists' Watches:
Romero Britto's "The Children
of the World".

Replicas

Six or seven years ago certain complicated wrist watch models from Movado achieved a high level of recognition – e.g. the Calendomatic, Chronograph, Chronoplan and Celestograph. Suddenly prices on the auction market increased considerably without any discernible reason. In any event, it certainly seemed a propitious time to reconsider the mechanical wrist watch.

These *two* factors induced Movado to launch a series of models with mechanical movements, after having produced quartz wrist watches almost exclusively for the preceding few years. The classics of the forties and fifties were not to be copied exactly, but the external appearance was to be very similar to the originals. These watches are encased in 18 ct gold, steel or steel/gold combinations. The series is named "Collection 1881" and thus commemorates the centenary of the company's foundation by Achille Ditesheim.

The first three models of this series were presented at the Basel Fair in 1990. There was an "Automatic" with the date in an aperture, a slim rectangular wrist watch with small seconds, in the style of the Curviplan with a manually wound Unitas Calibre 6490 N movement, and a perpetual calendar with phases of the moon as the top of the range model. Since then other models have been added: a beautiful Calendomatic, which is close to being an exact copy of the classical original with an ETA Calibre 2892-2 movement, a watch with a central date hand, small seconds and a small 24-hour dial at the 12, and another that is reminiscent of the Chronograph model.

251 a–d
Wrist watch models from
"Collection 1881".

The reproductions

A further series of Movado models consists of exactly detailed reproductions of well-known older wrist watches, fitted with mechanical movements and produced in limited numbers.

The first of these is doubtless also the most beautiful of the reproductions: it is the **First World War Soldier's Watch,** which came out in a limited edition of 250 pieces in 1993. It has an 18 ct yellow gold case, which makes it more of a de luxe reproduction of a soldier's watch, with a 17-jewel ETA mechanical movement. The watch is presented with a reproduction of a Swiss soldier's leather case filled with various items to care for his kit (brushes, sewing gear, leather polish, etc.). One of the brushes has been replaced by a wooden box in which this handsome watch is displayed. The reproduction aspect is further enhanced by the design of the guarantee certificate that accompanies the watch, a reproduction of a Swiss soldier's paybook.

The second of these reproductions is a Museum Watch with Horwitt's dial, very much in the form and dimensions of the original series with domed glass, a gold-plate case and movement with manual winding. This series came out in 1994, on the 75th anniversary of the founding of the Bauhaus movement in 1919, and is therefore limited to just 1919 pieces. It was named the **Bauhaus Watch** and is presented in a wooden case that is painted in the typical Bauhaus colours, red, blue and yellow. This model, the original of which was designed by an artist, could also be included in the Artists' Collection.

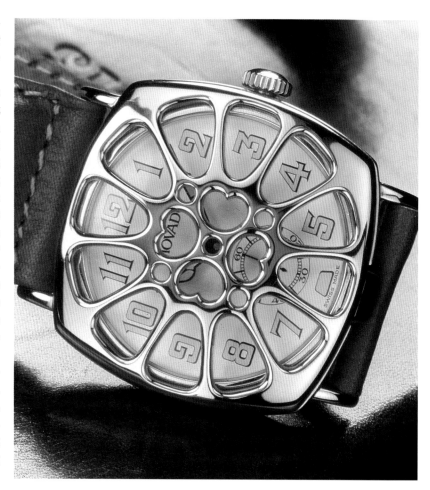

252 a, b
The reproduction of the Soldier's Watch from the First World War, made in 1993. The watch is cased in 18 ct gold and has a mechanical movement. View of Swiss Army cleaning outfit and the watch dial.

253
The "Bauhaus Watch" with Horwitt's Museum Dial, made in 1994. The case is gold plate (gold leaf on metal) fitted with a manually-wound mechanical movement. A general view of the watch in its original wooden presentation case.

Further models We have already written of the many versions of wrist watches with the Museum Dial currently produced; also of the fact that this dominates Movado's model programme today. Imitations of the original classical form, with only slight changes, play a rôle in many collections, for example in the Swiss Collection and the Classic Collection. In the **Black Sapphire Collection**, not only the dial but also the case and bracelet are black; in this watch the gold hands and dot at 12 are particularly striking. In the **Museum Olympian Watch** and the **Museum Sports Edition Watch**, this dial plays an equally decisive rôle design-wise, as in the **Museum Bracelet Watch** with its more festive gilded case and bracelet.

Besides these modern variations of Horwitt's classic dial, which one can hardly rank among the imitations, there are models similar to the classical watches which are not, however, intended to be reminiscent of specific watch types. For example the **Gentry Collection**, produced until quite recently, with finely styled dials in round or rectangular form, various calendar indications or a chronograph function. A very attractive chronograph model from the series has a black, white or blue dial with date display, hour and minute registers and an outer metal bezel with tachometer scale. The model is called "Chronograph 1950".

The **Swiss Collection** boasts a very pronounced futuristic design. It consists of various large rectangular wrist watches with strongly curved cases in which the upper and lower narrow sides of the case are optically disguised. The black dial, therefore, appears to be part of the continuous black strap, and stands out only because it is framed laterally by the broad gold-plate long sides of the case.

In the **Classic Collection** robust quartz watches are combined with bold metal bracelets in timeless and elegant designs. The Sport, Palaestrum, Emperium and Olympian models are particularly sporty with the crown integrated in the case; the Lyceum model is more elegant with a gilded case and bracelet. All the models are available in lady's and gentlemen's versions.

The lady's wrist watch series known as the **Minuet Collection** is very elegant with extravagantly formed metal bracelets and lavish dials, some inlaid with semi-precious stones or mother of pearl.

Let us end this chapter with a quotation from Gedalio Grinberg, which sheds a light on his attitude towards Movado and its acquisition:

"This is the beauty of this company. In addition to buying an outstanding design (the Museum Dial), we also happened to buy a very historically rich name which has made a major contribution to the history of watchmaking. To continue this tradition is as exciting as looking into the future."

Equally characteristic is the attitude of his son and designated successor Efraim:

"If you look back to the early 1900's, Movado was probably <u>the</u> brand – and we want to do the same thing, just on a bigger scale. Because to be successful today, we have to do things on a bigger scale. This is the beauty of Movado and this is the reason we're doing this book; going into its history. It's incredible. There's no other brand that has such a range of products and originality in product development. That's really what the company has been known for – for over a hundred years. The whole idea is to continue it."

254 a–c
Present-day range of
MOVADO quartz wrist
watches with the
Museum Dial.

The New.

1992   1912

MOVADO Switzerland

255 a, b
Advertisements showing the
Movado "Black Sapphire"
Collection.

256 a, b
Advertisements showing the
"Horizon" and "Sunwatch"
models.

Famous owners of Movado watches

Prominent personalities like to wear watches from companies which combine high quality with a high public level of fame. They could be divided into two main groups: those who take pleasure in wearing a proprietary watch that is prestigious or simply beautiful – this is frequently the case with actors and similar VIP's. Members of the second group value the reliability, quality and accuracy of a watch – the criteria that guide scientists or politicians. And thus we have found some personalities from public life who have owned and worn a Movado watch.

The most colourful is probably *King Boris III of Bulgaria*, the last monarch to reign over that country before the Communists seized power. The eldest son of King Ferdinand I, Boris became King in 1918 at the age of 24. In the following chaotic quarter of a century in the constantly restless Balkans, he managed with great diplomatic skill to manoeuvre between the Bolshevik and the National Socialist blocs. Boris remained an ally of the Axis powers without, however, renouncing diplomatic relations with the Soviet Union. With this seesaw policy he was able to keep his country out of the Second World War for a long time and to lead it to economic prosperity. His greatest achievement was preventing the deportation and murder of the Bulgarian Jews by the Nazis. On 14th August 1943, he was summoned to Hitler in the Wolf's Lair with a view to solving Bulgaria's Jewish problem in Hitler's way. But Boris stubbornly resisted Hitler's orders. Shortly after his return, Boris died suddenly and unexpectedly, from a heart attack, according to the official communiqué. It is still suspected that Hitler had him poisoned because of his intractability.

The pocket watch owned by King Boris was made by Movado in 1935. It is nothing like the ostentatious timepieces favoured by other crowned heads, being a simple and elegant, flat chronometer in a 14 ct gold case with movement Calibre 620N and the royal monogram in blue enamel on the case back. Boris's daughter Maria-Luisa, who lives in the USA, left it to her son, who sold it back to Movado in 1994.

An example of a politician's watch is the 18 ct gold calendar wrist watch of the Calendograph type with central seconds and movement Calibre 475 SC. It bears the following inscription on the case back: "Quelques citoyens helvétiques a *Winston Churchill* témoignage d'admiration et de reconnaisance – Septembre 1946" (Some Swiss citizens to Winston Churchill as a sign of their admiration and recognition). This was no doubt an attempt to pay tribute to the service of the great English politician and Prime Minister during the Second World War, irrespective of his defeat in the elections and subsequent resignation in July 1945. Unfortunately it is not known who these Swiss citizens were or whether Churchill ever wore the watch. It was offered for sale at a London auction in 1992.

The next famous owner of a Movado watch comes from the world of science and research: Professor *Auguste Piccard* (1884–1962). He was the Swiss physicist, stratosphere and deep-sea research scientist, who undertook the first ascent into the stratosphere in a balloon in 1931. Piccard was convinced that future air traffic would be affected at great heights which, due to lower air pressure, higher speeds and less fuel, would enable aircraft to travel greater distances – his theory was to be proven true later. On the wooden box of a large (81 mm diameter) deck watch in a steel case made by Movado, the dedication referring to one of his ascents reads: "Puisse ce chronomètre rendre toujours au Professeur Piccard les services qu'il a retiré durant sa glorieuse ascension dans la stratosphère 27.5.1931" (May this chronometer always be of service to Professor Piccard as on his famous ascent into the stratosphere 27.5.1931). On the case back we find the engraving "Au Professeur Piccard comme témoignage de vive admiration 27.5.1931" (For Professor Piccard as a sign of our lively admiration). It seems, therefore, that Professor Piccard had this watch with him on this balloon ascent, which took him to a height of ap-

257 a
Portrait of the Bulgarian Royal Family.

257 b–d
A flat 14 ct gold-cased Movado pocket watch for King Boris III of Bulgaria, made in 1935. Bi-coloured silvered dial with small seconds, signed "Chronomètre Movado". A large royal monogram in blue enamel is on the case back. Reference No. 0623156 47351. Nickel-plated 16¼''', Calibre 620 N movement with lever escapement, special patented precision regulator index, bimetallic compensation balance with Breguet balance spring, signed "Movado Factories", 15 jewels.

proximately 17,000 m. Unfortunately we have no further details about Piccard's chronometer or its donator. Perhaps it was one of those precision watches which were used for trials at one of the Swiss B.O.'s or at the Observatory in Neuchâtel or Teddington.

Here we should once more mention *Andy Warhol*, who was (see page 168) a keen watch collector and in whose collection there were several Movados. There were six Ermetos, among them one in a gold case with monogram, a Calendermeto, two Ermeto chronometers and a decorative red lacquered Baby Ermeto. Warhol's collection also included a small round coin watch with a St. Christopher on the case lid, a tonneau-shaped pocket watch and a calendar wrist watch. These watches were auctioned, together with other objects from Warhol's private possessions, 10,000 pieces in all, shortly after his death in April 1988 in a ten-day mammoth auction at Sotheby's in New York.

Two further prominent customers' watches come more from the sphere of entertainment. One is a beautiful rectangular gold wrist watch from about 1950 with a black dial and a signature "Cartier", with a Movado form movement Calibre 375. On the outside of the case back we find the engraving "*Alban Conway* with appreciation from *Moss Hart*". Moss Hart (1904 – 1961) was one of the best known American play and film scriptwriters. He was responsible for the script of the unforgettable Judy Garland film "A Star is Born", and in 1937, together with George S. Kaufmann, he was awarded the Pulitzer prize for the script of "You Can't Take It With You". In addition he wrote lyrics for Irving Berlin and Cole Porter. Moss Hart also worked as a stage director: in 1957 he won a Tony Award for the direction of "My Fair Lady" on Broadway. So much about the donator. We unfortunately know nothing about Alban Conway.

The second of these prominent customers' watches is a beautiful and tasteful 18 ct white gold wrist watch with a slightly tonneau-shaped rectangular case and lavishly worked lugs, or strap attachments. The case back is engraved with the inscription "Harry from Louella 1929". This watch was a gift from *Louella Parsons* to her second husband, the physicist Harry "Docky" Watson Martin. Louella Oettinger Parsons (1881 – 1972) started out as a reporter for the Chicago Tribune in 1910 and began writing one of the first film columns for the Chicago Record Herald in 1914. She was, beside Hedda Hopper, the best

known gossip columnist in Hollywood in the forties and fifties. The two women were not only famous but also feared, since they could destroy a hopeful acting career with one of their sharp-tongued articles.

258 a, b
Movado "Calendograph" cased in 18 ct gold, made for Winston Churchill in 1946, according to the dedication inscribed on the case back. Bi-coloured silvered dial with outer date chapter ring, month and day of the week shown in apertures, centre seconds, signed "Movado", Reference No. 4820. Nickel-plated 10¼''' Calibre 475 SC movement with lever escapement, 15 jewels.
(Privately owned).

259 c
Auguste Piccard

259 a, b
Large steel-cased deck
watch, made for Auguste
Piccard in 1931, according
to the dedication. The dial
with small seconds, signed
"Chronomètre Movado".
(Privately owned).

260
The 9 Movado watches in
the Andy Warhol collection.

262 a, b ▷
Rectangular 18 ct white gold-
cased wrist watch, given to
her husband by Louella Parsons in 1929, made in the
same year. The extravagant
case has flexible lugs. Bi-
coloured silvered dial with
small seconds, signed
"Movado". The case back is
engraved with the inscription
"Harry from Louella 1929",
Reference No. 740021.
Round 10¼''', Calibre 740
movement with lever escapement, 17 jewels.

261 a, b
Rectangular gold-cased Movado wrist watch given to Alban Conway by Moss Hart in 1950. Black dial with small seconds, signed "Cartier", Reference No. 43837 385531. Engraved inscription on the case back. Calibre 375 form movement with lever escapement, 17 jewels.

262 c
Caricature of Louella Parsons and Hedda Hopper.

Movado numbering and reference systems

Unlike many other watch manufacturers, Movado did not, from the beginning or from an early date, introduce a continuous numbering system for all the watches manufactured. Such a system, with movement numbers and entries for each movement in a movement book, enables the instant identification and dating of each watch. Over the years Movado used the four following numbering systems described below, which do not permit such identification and dating.

1. Up to 1966: a four or five-digit number engraved on the inside of the case back together with a six-digit serial number.
2. From 1966 to 1971: an eight or nine-digit number engraved on the inside of the case back.
3. From 1971 to 1983: a nine-digit number engraved on the outside of the case back.
4. Since 1983: a seven-digit number engraved on the outside of the case back.

Movement numbers on the movement itself appear only on chronometers that were intended to be tested by an observatory, or at a B.O., since these institutions required a movement identification number.

Re. 1 The original case reference system used up to 1966

This system was already adopted by Movado in the 19th century. If a watch has a four-digit number the case will be 18 ct gold. Other case materials are indicated by a fifth, code digit, preceeding the four-digit number, and can be identified via Table 2. The watch movement calibre may be identified using the remaining four digits in Table 1.

Example:

Reference No. 9038:   18 ct gold chronograph, Calibre 90 or 95 (according to Table 1)

Reference No. 19038:   stainless steel cased version of the above watch (according to Table 2)

Re. Table 1: With the introduction of further calibres the same four-digit numbers were reassigned to a different calibre (Table 1, Column "New").

The six-digit serial numbers were not always assigned in consecutive order. Very often, however, the first three digits give an indication of the movement calibre and the remaining numerals indicate the total quantity of movements of this calibre that had so far been manufactured.

Example:

Reference No. 6037:   according to Table, 1 Movement Calibre 550,

Reference No. 554371: this movement is the 4,371st of Calibre 550.

Around 1930 Movado began to stamp the calibre number on the movement itself. This number is normally located under the balance wheel. Occasionally, the calibre number had already been stamped on the dial side of the pillar plate.

Table 1   Case reference system and movement calibres (up to 1966)

Reference No.	Calibre Old	Calibre New
1–10	9M	–
11–20	760	685
21–99	780	5-6
100–300	Repeater	5-6
1001–1100	35	215
1101–1200	120	431
1201–1300	730	431/438
1301–1400	320	15-16
1401–1500	–	15-16
1501–1600	150M	15-16
1601–1800	150M	150 MN
1801–1900	510	–
1901–2000	510	190
2001–2100	50	335-337
2101–2200	50	47
2201–2300	440	230
2301–2400	45	45-46-48
2401–2500	Timer	–
2501–2600	250	245-246
2601–2700	250	55-56-57-58
2701–2800	75	55
2801–2900	75	75
2901–3000	–	–
3001–3100	25	345-347
3101–3200	25	348
3201–3400	105	105
3401–3500	340	165-166
3501–3600	350	167-169
3601–3700	360	421-425
3701–3800	370	423-427
3801–3900	375	–
3901–4000	375	–
4001–4100	400	–
4101–4400	65	65
4401–4450	420	65
4451–4500	480	65
4601–4700	423	365-368
4701–5000	470	470
5001–5100	530	531
5101–5200	580	–
5201–5300	600	538
5301–5400	–	205
5401–5500	580	205
5501–5600	800	800
5601–5700	500	–
5701–5800	500	–
5801–5900	720	–
5901–6000	–	–
6001–6100	550	600/608
6101–6400	–	220
6401–6500	820	220
6501–6600	821	220
6601–6700	For foreign countries	–
6701–6800	Ralco	–
6801–6900	–	–
6901–7000	540	540
7101–7200	620	–
7201–7300	640	135
7301–7500	620	–
7501–7600	750	–
7601–7800	660	–
7801–7900	680	575-578
7901–8000	575	575-578
8001–8100	–	–
8101–8400	125	125-127-128
8401–8600	115	115
8601–9000	–	260
9001–9100	90	90-95
9101–9200	–	395
9201–9300	–	380
9301–9400	–	388
9401–9500	–	–
9501–9600	950	–
9601–9700	910	–
9701–9800	950	–
9801–9900	–	–
9901–9999	–	900

Table 2   Numerals and letters for case material code (up to 1966)

Numeral/letter	Material
1	Stainless steel
2	Silver
3	Chromium plated
4	14 ct gold
5	Gold-plated case with stainless steel back
6	Complete case gold plated
7	
8	
9	9 ct gold
R	18 ct rose gold
B	Bi-coloured gold
G	18 ct white gold

Re. 2 The case reference system used from 1966 up to 1971

The new references appear in the form of 3 groups of numbers, each with 3 figures (Example: 389-215-518). Only the first group, which indicates the calibre, may have only one or two figures.

Explanation of the various groups:

1st group: Calibre XXX - xxx - xxx
   The calibre number used up to this time now appears in front of the reference number.

2nd group: Technical number xxx - XXX - xxx
   This part of the reference indicates by means of the first and second figures the case metal and by means of the third figure the type of case (see following table). 1st and 2nd figures:

10 platinum	57 white gold plated
21 18 ct yellow gold	61 40 mic. yellow gold
22 14 ct yellow gold	plated, steel back
24 18 ct pink gold	62 20 mic. yellow gold
25 14 ct pink gold	plated, steel back
27 18 ct white gold	64 40 mic. pink gold
28 14 ct white gold	plated, steel back
30 silver	65 20 mic. pink gold
48 yellow gold capped	plated, steel back
49 pink gold capped	67 white gold plated,
51 40 mic. yellow gold plate	steel back
52 20 mic. yellow gold plate	70 steel
54 40 mic. pink gold plate	80 chrome steel back
55 20 mic. pink gold plate	90 rough case

2nd group: 3rd figure

1 jewel watch with gold	leather strap
bracelet	4 round watch
2 watch with gold	5 fancy shape watch
bracelet	6 tournage
3 jewel watch with	7 Ermeto pocket watch

3rd group: Individual number xxx - xxx - XXX
   This number gives the serial number of an individual watch within its own calibre. The first figure indicates the type of crystal and whether the case is water-resistant or not.

from 001 to 299	ordinary crystal
from 300 to 399	sapphire crystal
from 400 to 499	facetted crystal
from 500 to 599	water-resistant crystal

Examples: previous G 2441 becomes:
48-271-041 (Calibre 48, 18 ct white gold, jewel watch with gold bracelet, ordinary crystal and individual number).
previous 19218 becomes:
380-705-518 (Calibre 380, steel case, special shape, water-resistant, and individual number).
See accompanying table: List of Old and New Calibre Numbers (1966–1971).

List of Old and New Calibre Numbers (1966–1971).

Previous Ancienne Anteriores Frühere	New Nouvelle Nuevas Neue	Previous Ancienne Anteriores Frühere	New Nouvelle Nuevas Neue	Previous Ancienne Anteriores Frühere	New Nouvelle Nuevas Neue
93	7-214-093	R 2504	246-244-004	G 2405	48-275-005
R 93	7-244-093	2507	246-215-007	2406	48-216-006
146	7-214-346	R 2507	246-245-007	R 2567	246-245-067
R 146	7-244-346	2508	246-217-008	2568	246-215-068
147	7-212-347	2509	246-215-009	R 2568	246-245-068
G 147	7-272-347	R 2509	246-245-009	2570	246-215-070
148	7-214-308	G 2509	246-275-009	R 2570	246-245-070
R 148	7-244-308	2511	246-215-011	G 2570	246-275-070
G 148	7-274-308	R 2511	246-245-011	G 2571	246-273-071
AL R 148	7-246-309	G 2511	246-275-011	2573	246-214-073
AL G 148	7-276-309	408	4-216-301	2579	246-212-079
192	7-214-313	R 408	4-246-301	2601	57-214-001
R 192	7-244-313	G 408	4-276-301	R 2601	57-244-001
199	7-213-300	G 413	7-271-113	2623	57-216-323
G 199	7-273-300	G 415	7-273-115	R 2623	57-246-323
G 223	7-273-223	G 418	4-276-001	G 2623	57-276-323
232	7-214-232	427	7-212-327	R 2680	57-245-080
R 232	7-244-232	G 438	4-271-008	2700	57-215-100
235	7-215-335	442	7-212-042	R 2715	57-245-015
R 235	7-245-335	G 442	7-272-042	2804	58-212-104
G 235	7-275-335	G 446	7-271-046	R 2406	48-246-006
247	7-214-303	448	7-214-311	G 2406	48-276-006
R 247	7-244-303	R 448	7-244-311	G 2420	48-276-020
281	7-213-302	R 455	7-242-355	2422	48-212-322
G 281	7-273-302	G 457	7-276-284	G 2422	48-272-322
G 298	7-276-298	458	4-212-302	2423	48-215-023
G 338	7-273-038	460	4-211-088	R 2423	48-245-023
348	7-216-348	G 460	4-271-088	2424	48-212-024
R 348	7-246-348	464	7-215-064	R 2424	48-242-024
G 348	7-276-348	469	7-212-069	G 2424	48-272-024
350	7-214-050	472	7-214-072	2432	48-215-332
R 350	7-244-050	G 483	7-271-020	R 2432	48-245-332
G 350	7-274-050	484	7-212-021	G 2432	48-275-332
G 356	7-273-056	G 484	7-272-021	AL G 2432	48-216-333
383	7-214-383	R 485	7-242-315	AL G 2432	48-276-333
R 383	7-244-383	488	7-216-028	2435	48-216-035
G 384	7-273-084	G 488	7-276-028	R 2435	48-246-035
393	7-214-095	492	7-212-305	G 2435	48-276-035
R 393	7-244-095	496	7-212-100	2436	48-216-036
G 393	7-274-095	1146	7-704-346	R 2436	48-246-036
G 399	7-271-099	1148	7-704-308	G 2436	48-276-036
G 400	7-216-281	2006	337-214-006	2437	48-212-037
G 403	4-273-003	2032	337-214-032	2438	48-212-038
2475	48-211-075	R 2032	337-244-032	R 2438	48-242-038
G 2475	48-271-075	2382	48-213-382	G 2438	48-272-038
2483	48-212-083	G 2382	48-273-382	G 2440	48-276-040
2487	48-212-087	2403	48-216-303	2441	48-211-041
2488	48-212-088	R 2403	48-246-303	G 2441	48-271-041
2492	48-212-092	G 2403	48-276-303	G 2450	48-276-050
2494	48-215-094	2405	48-215-005	G 2451	48-271-051
2504	246-214-004	R 2405	48-245-005	G 2452	48-276-052

Previous Ancienne Anteriores Frühere	New Nouvelle Nuevas Neue	Previous Ancienne Anteriores Frühere	New Nouvelle Nuevas Neue	Previous Ancienne Anteriores Frühere	New Nouvelle Nuevas Neue	Previous Ancienne Anteriores Frühere	New Nouvelle Nuevas Neue	Previous Ancienne Anteriores Frühere	New Nouvelle Nuevas Neue
2453	48-215-053	2535	246-215-035	2816	58-216-116	12844	58-705-009	G 5308	99-274-008
R 2453	48-245-053	R 2535	246-245-035	R 2816	58-246-116	12865	58-704-065	5309	99-215-009
2454	48-216-054	G 2535	246-275-035	G 2816	58-276-116	12877	58-704-077	19123	395-704-023
G 2454	48-276-054	2538	246-212-038	2846	58-212-046	12888	58-704-088	19141	395-704-541
2456	48-212-056	R 2538	246-242-038	R 2846	58-242-046	12998	58-704-498	19151	395-704-551
G 2456	48-272-056	2539	246-215-339	G 2846	58-272-046	13003	347-704-503	19161	395-704-561
2457	48-212-357	R 2539	246-245-339	2865	58-214-065	3601	425-214-501	19191	395-704-591
G 2457	48-272-357	2541	246-215-041	R 2865	58-244-065	R 3601	425-244-501	19204	380-705-004
G 2460	48-271-060	R 2541	246-245-041	2884	53-215-084	3603	425-214-003	19218	380-705-518
2463	48-212-363	2542	246-212-042	R 2884	53-245-084	R 3603	425-244-003	19230	380-704-530
2464	48-212-064	2543	246-212-043	2885	58-214-085	3604	425-214-304	19241	380-704-541
2465	48-212-065	2545	246-212-045	R 2885	58-244-085	R 3604	425-244-304	19261	380-704-561
G 2465	48-272-065	R 2545	246-242-045	2888	58-214-088	3605	425-215-005	19304	388-705-004
G 2961	58-273-361	G 2547	246-271-047	R 2888	58-244-088	R 3605	425-245-005	19318	389-705-518
2962	58-212-062	G 2548	246-272-048	2891	53-216-091	3606	425-215-006	19319	388-705-519
2963	53-215-063	2549	246-212-049	R 2891	53-246-091	R 3606	425-245-006	19323	388-704-023
2966	58-214-020	2555	246-215-055	G 2891	53-276-091	3610	425-214-010	19330	388-704-530
2986	58-214-086	R 2555	246-245-055	2894	58-214-394	R 3610	425-244-010	19341	388-704-541
R 2986	58-244-086	G 2555	246-275-055	R 2896	58-245-096	3612	425-212-012	19361	388-704-561
G 2994	53-273-094	2557	246-215-057	2898	58-214-398	3705	427-214-505	6148	7-514-308
3006	347-244-006	G 2557	246-245-057	R 2898	58-244-398	3711	427-214-511	R 6148	7-544-308
R 3006	347-244-006	G 2560	246-275-060	2911	53-216-111	R 3711	427-244-511	7229	135-214-029
3266	105-217-066	G 2562	246-276-062	R 2911	53-246-111	3712	427-212-512	R 7229	135-244-029
G 3266	105-277-066	2563	246-211-063	G 2911	53-276-111	3718	427-215-518	7975	578-217-071
3408	166-214-008	2565	246-212-065	2914	53-215-114	R 3718	427-245-518	9068	95-214-568
R 3408	166-244-008	G 2565	246-272-065	R 2914	53-245-114	3771	427-214-571	R 9068	95-244-568
3410	166-214-010	2567	246-215-067	R 2915	58-245-015	R 3771	427-244-571	9101	395-214-001
R 3410	166-244-010	R 5309	99-245-009	2927	58-212-027	5004	531-215-004	R 9101	395-244-001
3421	166-214-521	G 5309	99-275-009	G 2927	58-272-027	5102	531-214-102	G 9101	395-274-001
3471	169-214-571	G 5349	99-273-049	2942	58-214-342	R 5102	531-244-102	9103	395-214-003
R 3471	169-214-001	H 5353	205-245-053	R 2942	58-244-342	5104	531-215-104	R 9103	395-244-003
2518	246-215-018	G 5358	99-273-05	G 2942	58-274-342	5123	531-214-123	9104	395-215-004
R 2518	246-245-018	5369	99-215-069	2944	58-215-044	R 5123	531-244-123	R 9104	395-245-004
G 2521	246-273-021	G 5369	99-275-069	2948	58-212-048	5141	531-214-542	9106	395-214-506
2524	246-214-324	5371	202-217-071	G 2948	58-272-048	5179	531-214-579	R 9106	395-244-506
R 2524	246-244-324	R 5371	202-247-071	G 2949	58-272-048	R 5179	531-244-579	9107	395-214-507
2525	246-214-025	G 5371	202-277-071	G 2952	58-273-352	5204	538-215-004	R 9107	395-244-507
R 2525	246-244-025	5373	202-214-073	R 2953	58-245-053	R 5204	538-245-004	9115	395-214-015
G 2525	246-274-025	5393	7-624-095	2959	53-215-359	5223	538-214-023	R 9115	395-244-015
AL G 2525	246-216-026	R 5393	7-654-093	2961	58-213-361	R 5223	538-244-023	9118	395-215-518
AL R 2525	246-246-026	6012	608-214-012	9361	388-214-561	5231	538-214-531	R 9118	395-245-518
AL G 2525	246-276-026	R 6012	608-244-012	R 9361	388-244-561	R 5231	538-244-531	9122	393-215-022
2526	246-215-006	6014	608-215-014	9379	388-214-579	5232	7-624-232	R 9122	393-245-022
R 2526	246-245-006	R 6014	606-245-014	R 9379	388-244-579	R 5232	7-654-232	9141	395-214-541
2527	246-214-027	G 2807	58-275-107	11531	20-704-501	5261	538-214-561	R 9141	395-244-541
R 2527	246-244-027	2810	58-215-010	12511	246-705-011	R 5261	538-244-561	9151	395-214-551
G 2527	246-274-027	R 2810	58-245-010	12512	246-704-512	5279	538-214-579	R 9151	395-244-551
2528	246-214-028	AL G 2810	58-216-011	12526	246-705-006	R 5279	538-244-579	9204	380-215-004
R 2528	246-244-028	AL R 2810	58-246-011	12527	246-704-027	B G 5283	538-244-579	R 9204	380-245-004
2529	246-215-029	AL G 2810	58-276-011	12751	57-704-551	5308	99-214-008	9218	380-215-518
R 2529	246-245-029	2813	58-216-013	12803	58-704-303	R 5308	99-244-008	R 9218	380-245-5186

Previous Ancienne Anteriores Frühere	New Nouvelle Nuevas Neue	Previous Ancienne Anteriores Frühere	New Nouvelle Nuevas Neue	Previous Ancienne Anteriores Frühere	New Nouvelle Nuevas Neue	Previous Ancienne Anteriores Frühere	New Nouvelle Nuevas Neue	Previous Ancienne Anteriores Frühere	New Nouvelle Nuevas Neue	Previous Ancienne Anteriores Frühere	New Nouvelle Nuevas Neue
9230	380-214-530	15161	531-704-561	32664	246-805-054	R 59123	395-654-023	55261	538-614-561		
R 9241	380-244-541	15163	531-704-163	32556	246-805-056	59141	395-614-541	R 55261	538-644-561		
9301	388-214-001	15164	531-704-164	32798	57-804-498	R 59141	395-644-541	55279	538-614-579		
R 9301	388-244-001	15179	531-704-579	32894	58-804-494	59151	395-614-551	R 55279	538-644-579		
9304	388-215-004	15191	531-704-591	32908	58-804-008	R 59151	395-644-551	55291	538-614-591		
R 9304	388-245-004	15204	538-705-004	32950	58-805-050	59161	395-614-561	R 55291	538-644-591		
9318	389-215-518	15213	538-704-013	32964	58-804-064	R 59161	395-644-561	55341	202-624-041		
R 9318	389-245-518	15241	538-704-541	32974	58-805-074	52946	58-624-446	55345	202-614-545		
9322	388-215-022	15261	538-704-561	32998	58-804-498	R 52946	58-654-446	R 55345	202-644-545		
R 9322	388-245-022	15279	538-704-579	33007	347-804-010	52950	58-625-050	55352	202-624-052		
9328	389-212-528	15291	538-704-591	33023	347-804-023	R 52950	58-655-050	R 55353	205-655-053		
R 9328	389-242-528	15343	202-704-043	33407	166-804-007	52964	58-624-064	56011	608-614-511		
9330	388-214-530	15352	202-704-052	33602	425-805-002	R 52964	58-654-064	R 56011	608-644-511		
R 9330	388-244-530	15353	202-705-053	35164	531-804-164	52966	58-624-020	56014	608-625-014		
9341	388-214-541	15371	202-707-071	35353	205-805-053	52974	58-625-074	R 56014	608-655-014		
R 9341	388-244-541	16011	608-704-511	37204	135-804-003	R 52974	58-655-074	R 57204	135-654-004		
52865	59-624-065	16013	608-704-513	39901	895-807-001	52982	58-625-022	57220	135-624-020		
R 52865	58-654-065	16014	608-705-014	51531	20-614-501	R 52982	58-655-022	R 57220	135-654-020		
52884	53-625-084	16015	608-484-513	R 51531	20-644-501	52998	58-624-498	59191	395-614-591		
R 52884	53-655-084	16016	608-705-516	52032	337-624-032	R 52998	58-654-498	R 59191	395-644-591		
52888	58-624-088	17220	135-804-020	R 52032	337-654-032	53007	347-624-007	59218	380-615-518		
R 52888	58-654-088	17223	135-704-523	52511	246-625-011	R 53007	347-654-007	R 59218	380-645-518		
52894	58-624-494	19038	95-704-538	R 52511	246-655-011	53010	347-624-010	59241	380-614-541		
52908	58-624-008	19041	95-704-541	R 52516	246-655-016	R 53010	347-654-010	R 59241	380-644-541		
R 52908	58-654-008	19068	95-704-568	52526	246-625-006	53023	347-624-023	59304	388-625-004		
52940	58-625-440	19104	395-705-004	R 52526	246-655-006	R 53023	347-654-023	R 59304	388-655-004		
R 52940	58-655-440	19106	395-704-506	52527	246-624-027	53079	347-614-579	59318	389-615-518		
52944	58-625-044	19107	395-704-507	R 52527	246-654-027	R 53079	347-644-579	R 59318	389-645-518		
R 52944	58-655-044	19118	395-705-518	52533	246-624-033	53091	347-614-591	59323	388-624-023		
52945	58-624-445	53602	425-625-002	R 52533	246-654-033	R 53091	347-644-591	R 59323	388-654-023		
R 52945	58-654-445	R 53602	425-655-002	52553	246-625-053	53102	348-624-002	59330	388-614-530		
G 52945	58-674-445	53711	427-614-511	R 52553	246-655-053	R 53102	348-654-002	R 59330	388-644-530		
13007	347-704-007	R 53711	427-644-511	52554	246-625-054	53407	166-624-007	59341	388-614-541		
13010	347-704-010	53718	427-615-518	R 52554	246-655-054	R 53407	166-654-007	R 59341	388-644-541		
13023	347-704-023	R 53718	427-645-518	R 52556	246-655-056	53431	166-614-531	59361	388-614-561		
13089	347-704-589	53771	427-614-571	52556	246-625-056	R 53431	166-644-531	R 59361	388-644-561		
13091	347-704-591	R 53771	427-644-571	52680	57-625-080	53471	169-614-571	59379	388-614-579		
13101	348-704-501	55004	531-625-004	R 52680	57-655-080	R 53471	169-644-571	R 59379	388-644-579		
13102	348-704-002	R 55004	531-655-004	52798	57-624-498	R 55164	531-654-164	59901	895-507-001		
13421	166-704-521	55101	531-614-501	R 52798	57-654-498	55179	531-614-579	62603	57-514-303		
13471	169-704-571	R 55101	531-644-501	52844	58-625-009	R 55179	531-644-579	62698	57-514-498		
13509	169-704-509	55161	531-614-561	R 52844	58-655-009	55191	531-614-591	62803	58-514-303		
13531	169-704-531	R 55161	531-644-561	59104	395-625-004	R 55191	531-644-591	R 62803	58-544-303		
13701	427-704-501	55163	531-624-163	R 59104	395-655-004	55203	538-614-503	62835	58-514-335		
13711	427-704-511	55164	531-624-164	59106	395-614-506	R 55203	538-644-503	62942	58-514-342		
13716	427-705-516	19379	388-704-579	R 59106	395-644-506	55204	538-625-004	R 62942	58-544-342		
13718	427-705-518	S 31655	159-807-061	59107	395-614-507	R 55204	538-655-004	G 62948	58-575-048		
13771	427-704-571	S 31655	159-507-061	R 59107	395-644-507	55213	538-624-013	R 62948	58-545-048		
15004	531-705-004	32032	337-804-032	59118	395-615-518	R 55213	538-654-013				
15011	531-704-011	32527	246-804-027	R 59118	395-645-518	55241	538-614-541				
15101	531-704-501	32553	246-805-053	59123	395-624-023	R 55241	538-644-541				

Re. 3 The case reference system used from 1971 up to 1983

This reference system, still used by Zenith to the present day, was introduced shortly before the take-over by the Zenith Radio Corp. Chicago, and consists of nine figures in three groups.

Example: 36.0510.545

Explanation of the various groups:
1st group: consists of two figures. The first figure indicates the case material.

0	steel	5 gold with precious stones
1	gold plated	6 gold case and bracelet
2	gold plated	7 set with diamonds
3	gold	8 tournage
4	gold plated, steel back	9 tournage

The second figure of the first group distinguishes the gold case

0 yellow gold
1 white gold
2 rose gold

2nd group: consists of four figures, whereby the first three indicate the case model and the fourth the dial.

3rd group: this group is composed of three figures by which the movement calibre may be identified (see following list of calibres).

Movado-Zenith List of Calibres 1971–1983

Calibre No.	Calibre used	Size (in lignes)	Manu-facturer	Notes
011	511.1		Zenith	Clock
013	511.10		Zenith	Clock
025	990.1	10½	Longines	
034	3841	17¾		
035	117 T	17¾	Zenith	
036	3841	17¾		
040	8530	8¾	Citizen	
045	8510	8¾	Citizen	
050	279.001	5½	ETA	
051	279.001	5½	ETA	
053	979.001	5½	ETA	
054	978.001	5½	ETA	
055	977.001	5½	ETA	
065	101.001	4½	FHF	
084	952.111	11½		
085	951.111	7¾	Universal	
095	8710	12½	Citizen	
100	67	12½	Universal	Universal
104	956.031	7¾	ETA	
105	956.042	7¾	ETA	
106	956.112	7¾	ETA	
114	955.432	10½	ETA	
115	955.432	10½	ETA	
116	955.412	10½	ETA	
126	2542	11½	Zenith	
127	2542	11½	Zenith	For USA market
138	2552 PC	11½	Zenith	
139	2552 PC	11½	Zenith	17 jewels
150	1110	5½	Zenith	
154	1110	5½	Zenith	For USA market
155	1110	5½	Zenith	For USA market
157	1120	5½	Zenith	
160	1510	6¾	Zenith	
165	5340	5½	Omega	
172	2320	10½	Zenith	For Europ. & USA market
175	2310	10½	Zenith	For USA market
176	2320	10½	Zenith	For USA market
180	7001	10½	Peseux	
235	555.115	11½	ETA	
257	2562 C	11½	Zenith	
258	2562 C	11½	Zenith	For USA market
262	2540	11½	Zenith	
275	2552 C	11½	Zenith	
290	2562 PC	11½	Zenith	
305	5330	7¾	Omega	Omega 625
312	1724 C	7¾	Zenith	
315	950.001	7¾	ESA	
325	602.1	6 × 7¾	JLC	
335	603	7¾	JLC	
341	2824	11½	ETA	For USA market
345	2832	11½	ETA	
346	2837	11½	ETA	

Calibre No.	Calibre used	Size (in lignes)	Manufacturer	Notes
350	1730	$7^3/_4$	Zenith	For USA market
352	1730	$7^3/_4$	Zenith	For USA market
355	1740	$7^3/_4$	Zenith	
357	1730	$7^3/_4$	Zenith	17 jewels
365	2572 C	$11^1/_2$	Zenith	
366	2572 C	$11^1/_2$	Zenith	Without seconds hand
380	2572 PC	$11^1/_2$	Zenith	
382	2572 PC.E.	$11^1/_2$	Zenith	Economic version
415	3019 PHC	13	Zenith	
416	3019 PHC	13	Zenith	
421	B. 21		CEH	Beta 21
434	3019 PHC	13	Zenith	
435	3019 PHC	13	Zenith	For USA market
436	3019 PHF	13	Zenith	For USA market
437	3019 PHF	13	Zenith	
470		$11^1/_2$	Zenith	Quartz/analogue Version
471		$11^1/_2$	Zenith	Basis Calibre 470 – rectangular
480	2660	$7^3/_4$	ETA	
485	2671	$7^3/_4$	ETA	
490	2892	$12^1/_2$	ETA	
491	2892	$12^1/_2$	ETA	
500	9162-13	13	ESA	Mosaba
502	7	$5^3/_4$	Movado	
503	53	$5^1/_2 \times 6^1/_2$	Movado	
504	54	$5^1/_2 \times 6^1/_2$	Movado	
505	9164-13	13	ESA	Mosaba
506	57-6	$5^1/_2 \times 6^1/_2$	Movado	
507	57	$5^1/_2 \times 6^1/_2$	Movado	
508	58	$5^1/_2 \times 6^1/_2$	Movado	
510	9180	13	ESA	
515	9181	13	ESA	
517	169	$7^1/_4$	Movado	
518	166	$7^1/_4$	Movado	
522	425	$7^1/_4$	Movado	"Queenmatic"
523	427	$7^1/_4$	Movado	"Queenmatic"
525	421-1425	$7^1/_4$	Movado	"Queenmatic"
530	190	$8^3/_4$	Movado	
532	245	$8^3/_4$	Movado	Universal 800
533	246	$8^3/_4$	Movado	Universal 801
535	99	9	Piguet	Piguet 99/Rayville 58
540	150 MN	$10^1/_4$	Movado	
545	959.001	9	ETA	
547	335	$10^1/_2$	Movado	Universal 1005
548	336	$10^1/_2$	Movado	
549	337	$10^1/_2$	Movado	
550	345	$11^1/_2$	Movado	Universal 1105
551	346	$11^1/_2$	Movado	
552	347	$11^1/_2$	Movado	
553	348	$11^1/_2$	Movado	
555	290	$11^1/_2$	Movado	ETA 2391
555	999.001	$10^3/_4 \times 13$	ETA	Delirium – 1st version

Calibre No.	Calibre used	Size (in lignes)	Manufacturer	Notes
556	298	$11^1/_2$	Movado	ETA 2408
556	999.061	$10^3/_4 \times 13$	ETA	Delirium – 2nd version
571	387	$11^1/_4$	Movado	"Kingmatic S"
572	388	$11^1/_4$	Movado	"Kingmatic S"
573	389	$11^1/_4$	Movado	"Kingmatic S"
580	365	$11^1/_4$	Movado	
581	581.001	$8^3/_4$	ESA	
582	369	$11^1/_4$	Movado	
583	531	$11^3/_4$	Movado	"Kingmatic"
584	536	$11^3/_4$	Movado	"Kingmatic Chronometer"
585	538	$11^3/_4$	Movado	"Kingmatic Calendoplan"
587	600	$11^1/_2$	Movado	ETA 2471
588	608	$11^1/_2$	Movado	ETA 2472
590	965.101	$11^1/_2$	ESA	
600	2651	$7^3/_4$	ETA	
610	230	$12^1/_2$	Universal	Universal 66
619	135	$12^1/_2$	Movado	
620	540	17	Movado	
621	545	17	Movado	
632	47	$4^1/_2 \times 6$	Movado	
633	48	$4^1/_2 \times 6$	Movado	
634	48	$4^1/_2 \times 6$	Movado	For USA market
655	578	$5^3/_4 \times 7^1/_4$	Movado	"Ermeto Calendine"
660	105	$8^3/_4$	Movado	
675	404	$11^1/_4$	Movado	"Kingmatic HS 360"
685	407	$11^1/_4$	Movado	"Kingmatic HS 360" – "Presidents" model
686	407	$11^1/_4$	Movado	For USA market – "Presidents" model
690	408	$11^1/_4$	Movado	28-jewel "Kingmatic HS 360" calendar
692	408	$11^1/_4$	Movado	For USA market
695	409	$11^1/_4$	Movado	"Kingmatic HS 360"
700	405	$11^1/_4$	Movado	"Kingmatic HS 360" Day-Date Zenith
705	406	$11^1/_4$	Movado	"Kingmatic HS 360" Day-Date Movado
706	406	$11^1/_4$	Movado	"Kingmatic HS 360" Day-Date Movado
710	900	$11^1/_2$	Movado	"Ermetophone"
711	901	$11^1/_2$	Movado	"Ermetophone"

Re. 4 The case reference system used since 1983

The current reference system consists of a seven-digit number in three groups.

Example: 84. 40. 880.

Explanation of the various groups:

1st group: consists of two figures. They indicate the case material.

30	18 ct white gold
31	18 ct white gold plus diamonds
40	18 ct yellow gold
41	18 ct yellow gold plus diamonds
44	18 ct rose gold
70	14 ct yellow gold
71	14 ct yellow gold plus precious stones
80	Steel with decorated bezel
81	Steel with gold plated bezel
82	Metal white gold plated
84	Steel
85	Steel/gold bezel
86	Steel with gilded dots
87	Metal gold plated, steel back
88	Steel, yellow gold plated
90	Silver 925/1000
91	Gilded brass

2nd group: the third and fourth figures indicate the calibre number, from which the movement calibre may be identified (see the following lists of calibres).

3rd group: composed of three figures. The first (the fifth figure overall) indicates the case form.

2	Pendant	6	Television
3	Square	7	Oval
4	Rectangular	8	Round
5	Pocket watch	9	Special design

The second figure of the third group (the sixth figure overall) indicates the size of the case.

0	Lady's – mini	5	Gentleman's – mini
1	Lady's – small	6	Gentleman's – small
2	Lady's – standard	7	Gentleman's – medium
3	Lady's – medium	8	Gentleman's – standard
4	Lady's – large	9	Gentleman's – large

The third figure of the third group serves to define the case and is only relevant for manufacturing.

List of Calibres since 1983

MO	CALIBRE		REMARQUE	He	Mi	Se	Pe	V4	Fu	Qu	Jo	Mo	Ph	An	Ch	Cm	Cs	Cd	Ré	Rm	Batterie	Au	RU	H pi	H mouv	O ca	G 12H
	10½	7046		×	×	–	×	–	–	g	–	–	–	–	–	–	–	–	–	–	MANUEL			3.10	3.10	23.30	
	8¾	6480		×	×	–	*	–	–	g	–	–	–	–	–	–	–	–	–	–	MANUEL			3.95	3.95	19.40	
07	10½	255.241	CHRONO	×	×	×	–	–	–	g	–	–	–	–	–	×	–	–	–	–	395	30		3.50		23.30	
08	8¼	256.241	CHRONO	×	×	×	–	–	–	–	–	–	–	–	–	×	–	–	–	–	377	30		3.50		18.20	
09	5½	980.163	NORMFLAT	×	×	–	×	–	–	–	–	–	–	–	–	–	–	–	–	–	379	20		2.20		13.00	15.15
10	ROULET	3655A-77D		×	×	–	–	–	–	–	–	–	–	–	–	–	–	–	–	–	MR 8504					69.50	
	10½	737.3GP		×	×	×	–	–	–	–	–	–	×	–	–	–	–	–	–	–	371	27	11	2.75		23.30	
	10½	737.0GP		×	×	×	–	–	–	a	–	–	×	–	–	–	–	–	–	–	371	27	11	2.75		23.30	
	10½	737.2GP		×	×	×	–	–	–	g	–	–	–	–	–	–	–	–	–	–	371	27	11	2.75		23.30	
54	10½	255.411	FLAT III	×	×	×	–	–	–	g	–	–	–	–	–	–	–	–	–	–	373	27	6	2.10		23.30	
15	5½	1977		×	×	–	–	–	–	–	–	–	–	–	–	–	–	–	–	–	MANUEL		17	3.60		13.00	15.15
	11½	444.605		×	×	–	–	–	–	a	g	g	×	–	–	–	–	–	–	–	371	27	10	4.25		26.00	
	10½	255.511	FLAT III	×	×	×	–	–	–	g	–	–	–	–	–	–	–	–	–	–	373	36	6	2.10		23.30	
19	10½	255.441	FLAT III	×	×	–	–	–	–	g	–	–	–	–	–	–	–	–	–	–	373	27	6	2.10		23.30	
40	9	210.001	ELEGANCE	×	×	–	–	–	–	–	–	–	–	–	–	–	–	–	–	–	346	26	8	1.40	0.98	20.00	
40	9	210.001	ELEGANCE	×	×	–	–	–	–	–	–	–	–	–	–	–	–	–	–	–	341	32	8	1.60	0.98	20.00	
40	9	210.001	ELEGANCE	×	×	–	–	–	–	–	–	–	–	–	–	–	–	–	–	–	44	17	8	1.20	0.98	20.00	
21	9	210.011	ELEGANCE	×	×	–	–	–	–	–	–	–	–	–	–	–	–	–	–	–	341	36	8	2.00		20.40	
21	9	210.011	ELEGANCE	×	×	–	–	–	–	–	–	–	–	–	–	–	–	–	–	–	44	17	8	1.60		20.40	
42	9	210.011	ELEGANCE	×	×	–	–	–	–	g	–	–	–	–	–	–	–	–	–	–	341	36	8	2.00		20.40	
25	5½	976.001	FLAT III	×	×	–	–	–	–	–	–	–	–	–	–	–	–	–	–	–	321	24	7	1.95		13.00	15.15
35	10½	7001		×	×	–	*	–	–	–	–	–	–	–	–	–	–	–	–	–	MANUEL		17	2.50		23.30	
36	7¾	956.112	FLAT II	×	×	×	–	–	–	g	–	–	–	–	–	–	–	–	–	–	364	18	7	2.60	2.50	17.20	
36	7¾	956.112	FLAT II	×	×	×	–	–	–	g	–	–	–	–	–	–	–	–	–	–	377	26	7	3.10	2.50	17.20	
16	6¾	280.001	FLAT III	×	×	–	–	–	–	–	–	–	–	–	–	–	–	–	–	–	321	30	5	2.00	1.80	9.00	15.15
33	8¾	256.041	FLAT III	×	×	–	–	–	–	g	–	–	–	–	–	–	–	–	–	–	315	32	5	2.10		18.20	
44	8¾	956.431	FLAT II	×	×	–	–	–	–	–	–	–	–	–	–	–	–	–	–	–	364	36	7	2.50		18.20	
45	7¾	956.032	FLAT II	×	×	–	–	–	–	–	–	–	–	–	–	–	–	–	–	–	364	36	7	2.50		17.20	
46	8¼	256.031	FLAT III	×	×	–	–	–	–	–	–	–	–	–	–	–	–	–	–	–	315	32	5	2.10		18.20	
47	10½	955.422	FLAT II	×	×	×	–	–	–	g	g	–	–	–	–	–	–	–	–	–	371	27	7	2.50		23.30	
	4¼	201.101	ELEGANCE	×	×	–	–	–	–	–	–	–	–	–	–	–	–	–	–	–	321	32	2	2.50		9.90	
49	7¾	956.042	FLAT II	×	×	–	–	–	–	g	–	–	–	–	–	–	–	–	–	–	364	36	7	2.50		17.20	
	9	21		×	×	–	–	–	–	–	–	–	–	–	–	–	–	–	–	–	MANUEL		17	1.71		20.42	
32	8¼	256.111	FLAT III	×	×	×	–	–	–	g	–	–	–	–	–	–	–	–	–	–	315	20	6	2.10		18.20	
50	11½	255.121	FLAT III –	×	×	×	–	–	–	g	g	–	–	–	–	–	–	–	–	–	373	27	7	2.60		25.60	
56	11½	255.122	FLAT III	×	×	×	–	–	–	g	g	–	–	–	–	–	–	–	–	–	373	27	7	2.60		25.60	
59	11½	555.115	NORMLINE	×	×	×	–	–	–	g	–	–	–	–	–	–	–	–	–	–	371	30	1	2.75		25.60	
63	7¾	617 JLC		×	×	–	–	–	–	–	–	–	×	–	–	–	–	–	–	–	321	36		2.35		13.50	17,20
64	10½	955.432	FLAT II	×	×	–	–	–	–	–	–	–	–	–	–	–	–	–	–	–	371	36	7	2.50		23.30	
65	10½	955.412	FLAT II	×	×	×	–	–	–	g	–	–	–	–	–	–	–	–	–	–	371	27	7	2.50		23.30	
67	11½	888.444		×	×	–	–	–	–	a	g	g	×	–	–	–	–	–	–	–	AUTOM.			4.85		26.00	
	DELIR.	999.301	DELIRIUM –	×	×	–	–	–	–	–	–	–	–	–	–	–	–	–	–	–	333	15	4	1.68			
	DELIR.	999.001	DELIRIUM –	×	×	–	–	–	–	–	–	–	–	–	–	–	–	–	–	–	333	15	4	1.98			
51	11½	955.111	FLAT II	×	×	×	–	–	–	g	–	–	–	–	–	–	–	–	–	–	371	27	7	2.50		25.60	
	SCALA	999.351	DELIRIUM –	×	×	–	–	–	–	–	–	–	–	–	–	–	–	–	–	–	333		4	2.95			
	SCALA	999.251	DELIRIUM –	×	×	–	–	–	–	–	–	–	–	–	–	–	–	–	–	–	333		8	2.98			
	DELIR.	999.361	DELIRIUM –	×	×	–	–	–	–	–	–	–	–	–	–	–	–	–	–	–	321	18	8	2.58		20.00	23.60
	DELIR.	999.061	DELIRIUM –	×	×	–	–	–	–	–	–	–	–	–	–	–	–	–	–	–	321	18	8	2.58		24.50	29.60
52	11½	255.111	FLAT III	×	×	×	–	–	–	g	–	–	–	–	–	–	–	–	–	–	373	27	7	2.10		25.60	
	DELIR.	999.451	DELIRIUM –	×	×	–	–	–	–	–	–	–	–	–	–	–	–	–	–	–			0	0.98			
53	10½	955.411	FLAT II	×	×	×	–	–	–	g	–	–	–	–	–	–	–	–	–	–	371	27	7	2.50		23.30	
31	7¾	956.031	FLAT II	×	×	–	–	–	–	–	–	–	–	–	–	–	–	–	–	–	364	36	7	2.50		17.20	
90	5½	281.002	ELEGANCE	×	×	–	–	–	–	–	–	–	–	–	–	–	–	–	–	–	751	17	6	1.40		13.00	15.15
	12	1883	CHRONO	×	×	–	×	–	–	a	–	–	×	–	×	×	×	–	–	–	MANUEL		17	6.87		27.00	
94	7¾	556.115	NORMLINE –	×	×	×	–	–	–	g	–	–	–	–	–	–	–	–	–	–	364		1	2.75		17.20	
99	ROULET	PR 10WR		×	×	–	–	–	–	–	–	–	–	–	–	–	–	–	–	–	LR 1	12	0	11.30		50.00	
A0	11½	2892-2		×	×	–	–	–	–	–	–	–	–	–	–	–	–	–	–	–	AUTOM.		21	3.60		25.60	
A1	5½	980.003	NORMFLAT	×	×	–	–	–	–	–	–	–	–	–	–	–	–	–	–	–	317	24	7	2.20		13.00	15.15
A1	5½	980.003	NORMFLAT	×	×	–	–	–	–	–	–	–	–	–	–	–	–	–	–	–	379	36	7	2.50		13.00	15.15
A2	10½	955.432	FLAT II	×	×	–	–	–	–	–	–	–	–	–	–	–	–	–	–	–	371	36	7	2.50		23.30	

List of Calibres since 1983 (continued)

MO	CALIBRE		REMARQUE	He	Mi	Se	Pe	V4	Fu	Qu	Jo	Mo	Ph	An	Ch	Cm	Cs	Cd	Ré	Rm	Batterie	Au	RU	H pi	H mouv	O ca	G 12H
A3	7¾	956.114	FLAT II	×	×	×	–	–	–	g	–	–	–	–	–	–	–	–	–	–	364	18	7	2.50		17.20	
A4	11½	955.114	FLAT II	×	×	×	–	–	–	g	–	–	–	–	–	–	–	–	–	–	371	30	7	2.50		25.60	
A5	13¼	251.262	CHRONO	×	×	–	×	–	×	g	–	–	–	–	×	×	×	×	–	–	394	30	27	5.00		30.00	
A6	10½	955.424	FLAT II	×	×	×	–	–	–	g	g	–	–	–	–	–	–	–	–	–	371	27	7	3.00		23.30	
A7	13¼	283	CHRONO	×	×	–	×	–	–	g	–	–	–	–	×	×	×	–	–	–	AUTOM.		39	6.50		30.00	
A8	8¾	6580		×	×	–	×	–	–	g	–	–	–	–	–	–	–	–	–	–	MANUEL		17	3.95		19.40	
A9	11½	2892-2	MECALINE	×	×	×	–	–	–	g	–	–	–	–	–	–	–	–	–	–	AUTOM.		21		3.60	25.60	
B0	7¾	2671	MAS.OSC.OR	×	×	–	–	–	–	–	–	–	–	–	–	–	–	–	–	–	AUTOM.		25	4.80		17.20	
B1	11½	100-G	JAQUET-B-	×	×	×	–	–	–	a	g	g	×	–	–	–	–	–	–	–	AUTOM.		21	4.90		25.60	
B2	11½	876	JAQUET-B-	×	×	–	×	a	–	a	–	–	–	–	–	–	–	–	–	–	AUTOM.		21	5.20		25.60	
B3	11½	2890-9	ETA	×	×	–	–	–	–	a	a	a	×	×	–	–	–	–	–	–	AUTOM.		21	5.20		25.60	
B4	ROULET	260.039		×	×	–	–	–	–	–	–	–	–	–	–	–	–	–	–	–	LR 1	12	0	11.30		50.00	
B5	11½	926.301	NORMFLAT	×	×	–	–	–	–	–	–	–	–	–	–	–	–	–	×	–	399	40	7	4.60		25.60	
B6	ROULET	260.24		×	×	–	–	–	–	–	–	–	–	–	–	–	–	–	–	–	LR 1	12	0	11.30		50.00	
B7	ROULET	033		×	×	–	–	–	–	–	–	–	–	–	–	–	–	–	×	–	LZ1	12	0	14.80		50.00	
	9	959.001		×	×	–	–	–	–	–	–	–	–	–	–	–	–	–	–	–	315	24	6	1.63		20.20	
B9	7¾	2671	MECALINE	×	×	×	–	–	–	g	–	–	–	–	–	–	–	–	–	–	AUTOM.		25		4.80	17.20	
C1	5½	125	ISATRONIC	×	×	–	–	–	–	–	–	–	–	–	–	–	–	–	–	–	341	40	8	1.60	1.40	13.00	15.50
C2	11½	955.112	FLAT II	×	×	×	–	–	–	g	–	–	–	–	–	–	–	–	–	–	371	27	7	2.50		25.60	
C3	9	951	PIGUET	×	×	×	–	–	–	g	–	–	–	–	–	–	–	–	–	–	AUTOM.		19		3.25	20.40	
C4	8¾	1727	AS	×	×	–	×	–	–	–	–	–	–	–	–	–	–	–	–	–	MANUEL		17		3.55	19.40	
C5	13¼	251.272	CHRONO	×	×	–	×	–	×	g	–	–	–	–	×	×	×	–	–	–	394	30	27	4.60		30.00	
C6	7¾	127	ISATRONIC	×	×	–	–	–	–	–	–	–	–	–	–	–	–	–	–	–	341	40	8	1.60	1.40	17.20	
C7	8¾	1727	AS	×	×	–	–	–	–	–	–	–	–	–	–	–	–	–	–	–	MANUEL		17		3.55	19.40	
C8	8¾	2681	ETA	×	×	×	–	–	–	g	–	–	–	–	–	–	–	–	–	–	AUTOM.		25		4.80	19.40	
C9	8¾	2685	ETA	×	×	–	–	–	–	a	–	–	×	–	–	–	–	–	–	–	AUTOM.		25		5.35	19.40	
D1	6¾	579.005	ETA	×	×	–	–	–	–	–	–	–	–	–	–	–	–	–	–	–	362		6	2.75		15.20	17.80
D2	TRI-COMPAX		JAQUET-B.	×	×	–	×	–	–	–	–	–	–	–	×	×	–	–	–	–	AUTOM.		21	7.80		30.00	
D3	11½	897	JAQUET-B.	×	×	×	–	a	–	a	a	–	–	–	–	–	–	–	–	×	AUTOM.		21	5.70		25.60	
D4	11½	2892-2 BO	JAQUET-B.	×	×	×	–	–	–	g	–	–	–	–	–	–	–	–	–	–	AUTOM.		21	3.60		25.60	
D5	11½	700 8J	JAQUET-B.	×	×	–	–	–	–	–	–	–	–	–	–	–	–	–	–	×	MANUEL						
D6	8½	2688-JB	JAQUET-B.	×	×	×	–	–	–	a	–	–	–	–	–	–	–	–	–	–	AUTOM.		17		5.35	19.40	
D7	11½	2824-2	ETA	×	×	×	–	–	–	g	–	–	–	–	–	–	–	–	–	–	AUTOM.		25		4.60	25.60	
D8	11½	2801-2	ETA	×	×	×	–	–	–	–	–	–	–	–	–	–	–	–	–	–	MANUEL		21		3.35	25.60	
D9	8¾	2000	ETA	×	×	×	–	–	–	g	–	–	–	–	–	–	–	–	–	–	AUTOM.		20		3.60	19.40	
E1	8¾	956.412	ETA	×	×	×	–	–	–	g	–	–	–	–	–	–	–	–	–	–	397	26	7	3.10	2.50	19.40	
E1	8¾	956.412	ETA	×	×	×	–	–	–	g	–	–	–	–	–	–	–	–	–	–	362	17	7	2.60	2.50	19.40	
E2	10½	955.412	TWO HANDS	×	×	–	–	–	–	g	–	–	–	–	–	–	–	–	–	–	371	27	7	2.50		23.30	
E2	10½	955.412	TWO HANDS	×	×	–	–	–	–	g	–	–	–	–	–	–	–	–	–	–	395	48	7	3.12		23.30	
E3	7¾	956.112	TWO HANDS	×	×	–	–	–	–	g	–	–	–	–	–	–	–	–	–	–	364	18	7	2.60		17.20	
E3	7¾	956.112	TWO HANDS	×	×	–	–	–	–	g	–	–	–	–	–	–	–	–	–	–	377	26	7	3.10		17.20	
E4	6¾x8	902.001	ETA	×	×	–	–	–	–	–	–	–	–	–	–	–	–	–	–	–	364	34	4	2.35		15.30	18.20
E5	6¾x8	802.101	ART WATCH	×	×	×	–	–	–	–	–	–	–	–	–	–	–	–	–	–	377	40	0	2.95		15.30	18.20
E6	8¾	956.412	ETA	×	×	–	–	–	–	g	–	–	–	–	–	–	–	–	–	–	397	26	7	3.10	2.50	19.40	
E7	11½	955.112	ETA	×	×	–	–	–	–	g	–	–	–	–	–	–	–	–	–	–	395	48	7	3.12	2.50	25.60	
E8	7¾	127-1570	ISA	×	×	–	–	–	–	–	–	–	–	–	–	–	–	–	–	–	–	–	8	3.12	1.40	17.20	
E9	5½	980.153	ETA	×	×	–	×	–	–	–	–	–	–	–	–	–	–	–	–	–	379	20	11	2.55	2.20	13.00	15.15

Description of different Abbreviations:

He:	Aiguille des heures	Ch:	Compteur des heures
Mi:	Aiguille des minutes	Cm:	Compteur des minutes
Se:	Aiguille des secondes	Cs:	Compteur des secondes
Pe:	Aiguille petite seconde	Cd:	Compteur des 1/10 de sec.
V4:	Aiguille 24 heures	Ré:	Révell
Fu:	Fuseau horaire	Rm:	Réserve de marche
Qu:	Quantième	a:	par aiguille
Jo:	Jour de la semaine	g:	par guichet
Mo:	Mois	d:	affichage digital
Ph:	Phase de lune	*:	avec ou sans petite seconde
An:	Année		

List of all Movado mechanical calibres from 1910 to 1980

The following list contains, according to availability, all calibres used by Movado in the time span indicated. However, there are some gaps. For the complete list of calibres used between 1971 and 1983 see Appendix 1 (Movado-Zenith list of calibres).

These calibre lists were prepared by Bernhard Stoeber of Lyndhurst, and I express my sincere thanks to him.

Calibre	Diameter (mm)	Thickness (mm)	Functions	Vibs/ hour	Jewels	Year of introduction	End of production	Notes
3 M	8,90×20,80	3,50		18 000	17	ca. 1930	ca. 1935	
4	13,40	3,25		21 600	17	ca. 1960	1970	Back-wind
5	13,40	3,25		18 000	17	ca. 1950	ca. 1960	
5 M	27,00			18 000	15	ca. 1940	ca. 1950	Tavannes Calibre 214
6	13,40	3,25		21 600	17	ca. 1960	1961	Basis Calibre 5
7	13,40	3,25		21 600	17	ca. 1965	ca. 1975	Basis Calibre 6, balance without screws
9 M	9,00×18,00			18 000	17	1932	1950	
11	11,50×19,50	3,30		18 000	17	1926	1940	Replaced by Calibre 575
15	16,50	3,50		18 000	17	1947	1958	
16	16,50	3,50		21 600	17	1958	1960	Basis Calibre 15
17	16,50	3,50	SCI	18 000	17	1950	1958	
19	16,50	3,50	SCI	21 600	17	1958	1960	Basis Calibre 17
20	15,30	4,08	SCD	21 600	17	1966	1970	FHF Ebauche, Calibre 57
25	14,00×18,00	3,05		18 000	15	1938	ca. 1945	
27	14,00×18,00	3,85	SCI	18 000	15	1940	ca. 1945	Basis Calibre 25
28	14,00×18,00	4,45	CLD QG	18 000	15	1940	ca. 1955	Basis Calibre 25
35	11,30×20,40	3,30		18 000	15	ca. 1935	ca. 1945	
45	10,25×13,70	3,50		18 000	17	ca. 1960	ca. 1961	
46	10,25×13,70	3,50		21 600	17	ca. 1960	1961	Basis Calibre 45
47	10,25×13,70	3,50		21 600	17	ca. 1960	ca. 1970	Basis Calibre 45
48	10,25×13,70	3,50		21 600	17	ca. 1960	1971	Basis Calibre 46, balance without screws
50 SP	17,00	3,05		18 000	15	1924	ca. 1950	Replaced by Calibre 15
53	12,20×14,75	3,35		21 600	17	1966	1970	Also diam. 15.35 mm, back-wind
54	12,20×14,75	3,35		21 600	17	ca. 1960	1966	Back-wind
55	12,20×14,75	3,35		18 000	17	1956	1958	Kif shock absorbers
56	12,20×14,75	3,35		21 600	17	1958	1966	Kif shock absorbers
57	12,20×14,75	3,35		21 600	17	1966	1967	Also diam. 15.70 mm, balance without screws
58	12,20×14,75	3,35		21 600	17	1967	1971	Also diam. 15.35 mm, balance without screws
65	11,50×16,00	3,60		18 000	15	ca. 1935	1957	
75	25,50			18 000	15	ca. 1930	ca. 1960	From approx. 1940 as Calibre 75 SC with centre secs.
85	19,70		SCD	18 000	17	ca. 1960	ca. 1950	Vulcain Calibre 406 "Alarm"
90 M	26,60	5,60	CHR 60M	18 000	17	1938	ca. 1965	With chronograph module
95 M	26,60	6,75	CHR 60M 12H	18 000	17	1939	ca. 1970	With chronograph module
99	20,40	1,77		18 000	17	ca. 1960	1975	Piguet Calibre 99/Rayville 58
105	19,50	3,50		18 000	15	ca. 1910	ca. 1950	
105 SC	19,50	4,40	SCI	18 000	15	ca. 1930	ca. 1940	From approx. 1930 also for Ermetolux
107	19,50	4,40	SCI	18 000	15	ca. 1930	ca. 1950	Basis Calibre 105
115	26,50	4,45	AUT	18 000	15	ca. 1950	1959	For "Futuramic"
116	26,50	4,45	AUT	18 000	21	ca. 1950	ca. 1959	Basis Calibre 115, for chronometers
118	26,50	5,50	AUT CLD JG	18 000	17	ca. 1950	1959	Basis Calibre 115-Calendolux, day + date in apertures
120				18 000	15	ca. 1910	ca. 1930	Alarm
122	28,00	4,85	CLD	18 000	17	ca. 1950	ca. 1960	Basis Calibre 125-Calendoplan, without date corrector
123	28,00	5,65	SCI CLD	18 000	17	ca. 1950	ca. 1960	Basis Calibre 125-Calendoplan, Incabloc, without date corrector
125	28,00	3,90		18 000	17	1946	1961	

Calibre	Diameter (mm)	Thickness (mm)	Functions	Vibs/ hour	Jewels	Year of introduction	End of production	Notes
126	28,00	3,90		18 000	21	1946	ca. 1960	Basis Calibre 125, for chronometers
127	28,00	4,80	SCI	18 000	17	1952	1960	Incabloc shock absorbers
128	28,00	4,85	CLD	18 000	17	ca. 1950	ca. 1965	Basis Calibre 125-Calendoplan, Incabloc, date corrector
128 SC	28,00	5,65	SCI CLD	18 000	17	ca. 1950	ca. 1960	Basis Calibre 125-Calendoplan, Incabloc, date corrector
129	28,00	4,65		18 000	17	ca. 1950	ca. 1965	Basis Cal. 125, "Polygraf" world time watch
135	28,00	3,98		21 600	17	ca. 1960	1970	
150 MN	22,90	4,25		18 000	17	ca. 1910	ca. 1975	
150 MN	22,90	4,25		18 000	17	1926	1954	"Ermeto" Non-Stop
155	22,90×24,00	5,60	QA JG MG PH	18 000	17	1948	1954	"Calendermeto" Non-Stop with calendar module
157	22,90	4,75	SCI	18 000	17	ca. 1935	ca. 1954	Basis Calibre 150, also for "Ermeto"
159	22,90	6,20	SCI CLD	18 000	17	1959	1970	"Ermetoscope"
165	16,50	4,90	AUT	18 000	17/30	ca. 1950	ca. 1960	
166	16,50	4,90		21 600	17/30	ca. 1960	1965	
167	16,50	5,60	AUT SCI	18 000	17/30	ca. 1950	ca. 1960	Basis Calibre 165
169	16,50	5,60	SCI	21 600	17/30	ca. 1960	1965	Basis Calibre 166
190	19,50	4,70	SCD	18 000	17	1951	ca. 1965	
205	22,80	3,35		18 000	17	1955	1970	Basis Calibre 115, but smaller, without automatic system
220	27,00	5,70	AUT SCI	18 000	17	ca. 1945	ca. 1950	"Tempomatic"
220 M	28,00	5,70	AUT SCI	18 000	17	ca. 1945	ca. 1950	Basis Calibre 220, "Tempomatic"
221	27,00	5,70	AUT SCI	18 000	17	ca. 1950	ca. 1950	"Tempomatic" with heavy metal rotor
221 A	28,80	5,70	AUT SCI	18 000	17	1952	ca. 1954	"Tempomatic" with heavy metal rotor, Incabloc
222	27,00	6,50	AUT SCI CLD QG	18 000	17	1952	1954	"Calendoplan Automatic", Incabloc
223	27,00	7,05	AUT SCI CLD QA JG MG	18 000	17	1950	ca. 1954	"Calendomatic", Incabloc
223 A	28,80	7,05	AUT SCI CLD QA JG MG	18 000	17	1952	1950	"Calendomatic", Incabloc
224	27,00	6,50	AUT SCI CLD QG	18 000	17	ca. 1952	ca. 1954	"Calendoplan Automatic" with date corrector
224 A	27,00	6,50	AUT SCI CLD QG	18 000	17	1953	1954	"Calendoplan Automatic" with date corrector
225	27,00	7,05	AUT SCI CLD QA JG MC	18 000	17	ca. 1948	ca. 1950	Basis Calibre 220 "Calendomatic"
225 M	28,80	7,05	AUT SCI CLD QA JG MC	18 000	17	ca. 1948	ca. 1950	Basis Calibre 220 M "Calendomatic"
226	28,80	5,70	AUT SCI	18 000	17	ca. 1945	ca. 1950	Basis Calibre 220 M "Tempomatic"
226 A	28,80	5,70	AUT SCI	18 000	17	ca. 1945	ca. 1950	"Tempomatic", Incabloc
228	28,80	7,05	AUT SCI CLD QA JG MC	18 000	17	ca. 1948	ca. 1950	"Calendomatic" with calendar module
228 A	28,80	7,05	AUT SCI CLD QA JG MC	18 000	17	ca. 1948	ca. 1950	"Calendomatic" with calendar module, Incabloc
230	28,00	2,50	AUT	19 800	25	1966	1970	Universal ébauche Calibre 66 "Gentleman"
2301/231	28,00	2,50	AUT CLD QG	19 800	25	ca. 1966	1970	Universal ébauche Calibre 66, improved automatic system
245	19,40	2,45		21 600	17	ca. 1960	1966	Universal ébauche Calibre 820
246	19,40	2,45		21 600	17	ca. 1965	1971	Universal ébauche Calibre 820

Calibre	Diameter (mm)	Thickness (mm)	Functions	Vibs/ hour	Jewels	Year of introduction	End of production	Notes
250 N	38,00/40,00			18 000	15	1920	ca. 1930	Very flat pocket watch
260	23,30	4,15	SCI	18 000	17	ca. 1945	ca. 1950	
260 M	23,30	4,15	SCI	18 000	17	ca. 1945	ca. 1950	Basis Calibre 260
261	23,30	4,15	SCI	18 000	17	ca. 1950	ca. 1952	
261 A	23,30	4,15	SCI	18 000	17	ca. 1952	1954	Basis Calibre 261, Incabloc, improved centre secs.
280	23,30	3,60	SCD	18 000	17	ca. 1960	1970	Eta ébauche Calibre 2390
290	25,60	3,60	SCD	18 000	17	ca. 1960	1970	Eta ébauche Calibre 2391
298	25,60	3,60	SCD CLD	18 000	17	ca. 1960	1971	Eta ébauche Calibre 2408
310 M	41,40		S	18 000	15	ca. 1930	ca. 1940	"Ermeto Pullman", 8-day watch, no alarm
331	25,60		S QG JG	18 000	17	1953	ca. 1954	"Calendolux", similar to Calibres 118/115
335	23,30	3,40	SCD	18 000	17	ca. 1960	1966	Universal ébauche Calibre 1005
336	23,30	3,40	SCD	18 000	17	ca. 1960	ca. 1966	Universal ébauche Cal., same as Cal. 337
345	25,60	3,40	SCD	18 000	17	ca. 1960	1966	Basis Calibre 335
346	25,60	3,40	SCD	18 000	17	ca. 1962	1967	Basis Calibre 336
347	25,60	3,40	SCD	18 000	17	ca. 1962	ca. 1967	Basis Calibre 346
348	25,60	3,80	SCD CLD	18 000	17	ca. 1962	ca. 1967	Basis Calibre 347
350	40,0 bzw. 43,0		S	18 000	21			Pocket chronometers, deck watches
360	50,0 bzw. 65,0		S	18 000	21			Pocket chronometers, deck watches
365	25,30	3,65	SCD	18 000	17	ca. 1958	1965	Same construction as Calibre 431, without automatic
368	25,30	4,00	SCD CLD	18 000	17	ca. 1958	1967	Basis Calibre 365
369	25,30	4,00	SCD CLD RDR	21 600	17	ca. 1958	1970	Basis Calibre 365
375	18,50×22,10	3,30		18 000	15	ca. 1940	ca. 1960	
377	18,50×22,10	3,80	SCI	18 000	15	ca. 1940	ca. 1960	Basis Calibre 375
380	25,30	3,88	AUT SCD	21 600	17/28	1966	1970	"Kingmatic S", also as chronometer
387	25,30	4,16	AUT SCD QG RDR	21 600	17/28	1966	1970	"Kingmatic S", replaced by Calibre 407
388	25,30	4,16	AUT SCD QG RDR QG RDR	21 600	17/28	1966	1970	"Kingmatic S", also as chronometer, replaced by 408
389	25,30	4,36	AUT SCD CLD	21 600	17/28	1966	1970	"Kingmatic S", also as chronometer, replaced by 409
395	25,80	3,98	AUT SCD	21 600	17/28	1962	1967	"Kingmatic S"
395 A	25,30	3,98	AUT SCD	21 600	17/28	1963	1967	"Kingmatic S", similar to 395 B-395 F
400	15,20×38,00	4,50	S oder ohne Sek.	18 000	15	1912	1917	"Polyplan", also as chronometer
404	25,30	5,25	AUT SCD CLD JG RDR	36 000	17/28	1968	1972	"Kingmatic HS 360"
405	25,30	5,25	AUT SCD CLD QG JG RDR	36 000	17/28	1968	1972	"Kingmatic HS 360"
407	25,30	4,25	AUT SCD CLD RDR	36 000	17/28	1968	1972	"Kingmatic HS 360"
408	25,30	4,25	SCD QG RDR AUT	36 000	17/28	1968	1972	"Kingmatic HS 360", similar to Calibre 409
420	30,00			18 000	15	1914	ca. 1917	"Soldier's Watch"
421	16,20	4,25	AUT	19 800	17/25	1965	1970	Universal ébauche "Queenmatic", replaced by Zenith Calibres 1724-1725
423	16,20	4,75	AUT SCI	19 800	28	1965	1970	Basis Calibre 421 "Queenmatic", replaced by Zenith 1725-1724
425	16,20	4,25	AUT	19 800	17/25	ca. 1965	ca. 1978	Universal ébauche "Queenmatic", replaced by Calibre 1425
427	16,20	4,75	AUT SCI	19 800	17/25	ca. 1965	ca. 1978	Basis Calibre 425 "Queenmatic", replaced by Calibre 1427
431 A	25,50	6,00	AUT SCD	18 000	17/28	1956	1960	"Rotor Automatic"

Calibre	Diameter (mm)	Thickness (mm)	Functions	Vibs/ hour	Jewels	Year of introduction	End of production	Notes
436	25,50	6,00	AUT SCD	18 000	17/28	ca. 1958	ca. 1960	"Rotor Automatic"
438	25,50		AUT SCD CLD	18 000	17/28	1959	1962	"Rotor Automatic"
440	15,40×27,00	3,60		18 000	15	1930	ca. 1940	"Novoplan"
443	15,40×27,00	4,05	SCI	18 000	15	ca. 1930	ca. 1940	Basis Calibre 440
470	23,10	3,25	S oder ohne Sek.	18 000	15/17	1927	1954	
473	23,10×24,00	5,50	CLD MG JG QA PH	18 000	15/17	1947	1954	"Celestograf" with calendar, see also Calibre 155
473 SC	23,10×24,00	6,05	SCI CLD MG JG QA	18 000	15/17	1947	1954	"Celestograf" with calendar, see also Calibre 155
475	23,10×28,00	4,55	CLD MG JG QA	18 000	15/17	1938	1954	"Calendograf" with calendar module
475 SC	23,10×28,00	5,10	SCI CLD MG JG QA	18 000	15/17	1938	1954	"Calendograf" with calendar module
477	23,10	3,90	SCI	18 000	15/17	1927	1954	
478	Z3,10	4,65	S CHR	18 000	17/23	1941	1954	Single-button chronograph, without hour register
479	23,10×24,00	6,05	SCI CLD MG JG QA	18 000	15/17	1938	1954	"Calendograf" with calendar (see Calibres 155 & Calibre 473 SC)
480	30,00			18 000	15	ca. 1914	ca. 1917	Similar to Calibre 420 but open-face layout
485	17,20	4,80	AUT SCD	21 600	17/23	ca. 1975	ca. 1980	Eta ébauche Calibre 2671
500	12,50×22,50			18 000	15	ca. 1910	ca. 1930	
510	20,00×27,00	4,25		18 000	17	1926	ca. 1940	"Curviplan"
530	25,50			18 000	15	1902	ca. 1950	
531	26,50	5,25	AUT SCD	21 600	17/28	1960	1967	"Kingmatic", similar to Calibre 536 for chronometers
538	26,50	5,25	AUT SCD CLD RDR	21 600	17/28			"Kingmatic"
540	38,00	3,70		18 000	17	1934	1966	
545	38,00	3,70		18 000	17	1966	1968	With Kif shock absorbers
550	15,40×25,30	4,00		18 000	15	ca. 1910	ca. 1925	
575	13,20×16,60	3,70		18 000	17	1947	1954	Also used for "Ermeto Baby"
578	13,20×16,60	4,30	CLD	18 000	17	ca. 1948	ca. 1975	"Ermeto Calendine"
579	13,20×16,60		CLD SCI	18 000	15	ca. 1948	1960	"Calendoplan Baby"
580	24,00		S	18 000	15	ca. 1910	ca. 1930	Hunter
600	24,80			18 000	15	ca. 1910	ca. 1930	
600	25,60	5,20	AUT	18 000	17/25	ca. 1960	1969	Eta ébauche Calibre 2471
608	25,60	5,20	AUT SCD CLD QG	18 000	17/25	ca. 1960	1970	Eta ébauche Calibre 2472
620 N	36,80 (17+18‴)		S	18 000	17	1910	ca. 1940	
627	36,80		SCI	18 000	17	ca. 1920	ca. 1940	
640	36,80	3,80	S	18 000	15	ca. 1910	ca. 1930	Hunter
660	38,00		S	18 000	15	ca. 1910	ca. 1930	Open face
680	36,80		S	18 000	15	ca. 1910	ca. 1930	Hunter
711	47,50			18 000	17	ca. 1935	ca. 1945	"Ermeto Pullman", Lemania Calibre 711, 8-day movement with alarm
730	19,50	3,40	S	18 000	17	ca. 1920	ca. 1930	Replaced by Calibre 105
750	12,50×19,50			18 000	15	ca. 1910	ca. 1930	
760	19,50		S	18 000	17	ca. 1910	ca. 1930	
780	19,50	4,00	S	18 000	17	ca. 1914	ca. 1930	
800 M	42,40/47,40	5,60		18 000	15	ca. 1910	ca. 1960	Pocket watch calibre
820	43,00			18 000	15	ca. 1910	ca. 1920	Hunter, as open face Calibre 821
895	42,30	10,70		18 000	15	ca. 1945	1970	"Ermeto Pullman", AS-Calibre 895, 8-day movement with alarm
900	25,60	5,75	SCD	18 000	17	1955	ca. 1978	"Ermetophone", AS-Calibre 1475 with alarm

Calibre	Diameter (mm)	Thickness (mm)	Functions	Vibs/ hour	Jewels	Year of introduction	End of production	Notes
901	25,60	5,75	SCD	18 000	17	1955	1978	"Ermetophone", AS-Calibre 1475 with alarm
950	21,10			18 000	15	1920	ca. 1930	
1005	23,30	3,40	SCD	18 000	17	ca. 1960	1966	Universal Calibre 1005 (see Calibre 335)
1105	25,60	3,40	SCD	18 000	17	ca. 1962	ca. 1966	Universal Calibre 1105 (see Calibre 345)
1110	11,50×13,70	3,30		21 600	17	ca. 1972	ca. 1978	Zenith ébauche Calibre 1110
1425	16,20	4,75	AUT	19 800	17/25	ca. 1965	ca. 1978	"Queenmatic", see Calibre 425
1427	16,20	4,75	AUT SCI	19 800	17/25	ca. 1965	ca. 1972	"Queenmatic", see Calibre 427
1724 C	17,20	5,00	AIJT SCD QG	19 800	17/25	ca. 1972	ca. 1974	Zenith Calibre 1724
1725	17,20	4,35	AUT	19 800	17/25	ca. 1972	ca. 1974	Zenith Calibre 1725
1730	17,20	2,70		21 600	19	ca. 1972	ca. 1978	Zenith Calibre 1730
1740	17,20	2,70		21 600	19	ca. 1972	ca. 1978	Zenith Cal. 1740 with Kif shock absorbers
2320	23,40	2,70		21 600	17	ca. 1972	ca. 1980	Zenith ébauche Calibre 2310
2542	25,60	4,07	SCD	21 600	17	ca. 1972	ca. 1978	Zenith ébauche Calibre 2542
2542 C	25,60	4,07	SCD QG	21 600	17	1970	1972	Zenith ébauche Calibre 2542 C
2552 PC	25,60	5,63	AUT SCD QG	21 600	23	ca. 1972	ca. 1978	Zenith Calibre 2552 PC, ball bearing rotor
2562 C	25,60		SCD QG	28.600	17	ca. 1972	ca. 1978	Zenith Calibre 2562 C
2562 PC	25,60	5,63	AUT SCD QG	28 800	23	ca. 1972	ca. 1978	Zenith Calibre 2562 PC, ball bearing rotor
2572 C	25,60	4,07	QG keine Sek.	21 600	17	1972	1978	"Museum Watch" with date in the dot
3019 PHC	30,00	6,50	AUT S QG RDR CHR	36 000	17/31	1969	1980	"El Primero", "Datron"
3019 PHF	30,00	7,55	AUT S QG JG MG PH RDR CHR	36 000	17/31	ca. 1970	1980	"Astronic"

Key to abbreviations:

A/h	Vibrations per hour
AUT	Automatic winding
CHR	Chronograph
CLD	Calendar, date indication
JG	Date indication in an aperture
M	Modification
MN	2nd modification
MG	Month indication in an aperture
PH	Moon phase display
QA	Date hand
QG	Date indication in an aperture
RDR	Date quick set
S	Small seconds
SC	Centre seconds
SCD	Direct centre seconds
SCI	Indirect centre seconds
SP	Flat balance spring
60 M	60-minute register
12 H	Hour register

263
Motorist's watch with a chromium-plated case with steel back, made in 1935. The dial signed "Movado", Case Reference No. 11769, Serial No. 182452, Calibre 150 MN movement, 15 jewels, 3 chatons, 4 adjustments.

Illustrations of Calibres 4 to 900

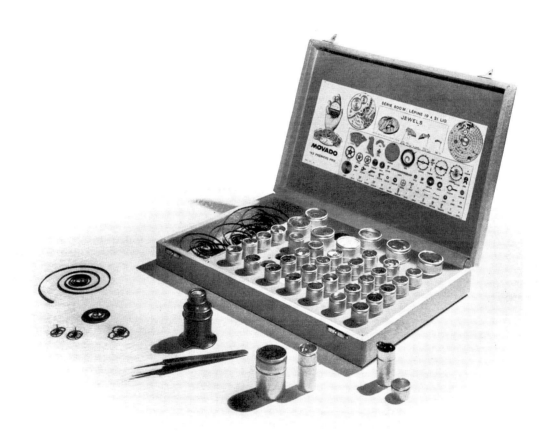

Cal. 4	**Cal. 5–6**	**Cal. 7**	**Cal. 15–16**	**Cal. 17–19**	**Cal. 20**
5¾′′′	4¾′′′	5¾′′′	7¼′′′	7¼′′′	6¾′′′
13,40 mm	13,40 mm	13,40 mm	16,50 mm	16,50 mm	15,30 mm

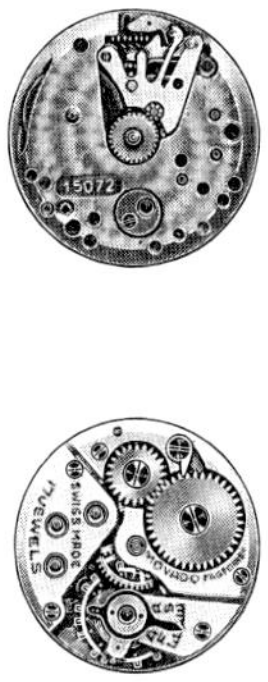

Cal. 45–46	**Cal. 48**	**Cal. 50 S. P.**	**Cal. 54**	**Cal. 55–56**	**Cal. 57**
4½ x 6′′′	4½′′′ x 6′′′	7½′′′	5½′′′ x 6½′′′	5½ x 6½′′′	5′′′ ½ x 6½′′′
10,25 x 13,70 mm	10,25 x 13,70 mm	17,00 mm	12,20 x 14,70 mm	12,20 x 14,70 mm	12,20 x 14,70 mm
			∅ 15,70 mm – 7′′′		∅ 15,70 mm – 7′′′
			Cal. 53		**Cal. 58**
			5½′′′ x 6½′′′		5′′′ ½ x 6½′′′
			12,20 x 14,75 mm		12,20 x 14,75 mm
			∅ 15,35 mm – 6¾′′′		∅ 15,35 mm – 6¾′′′

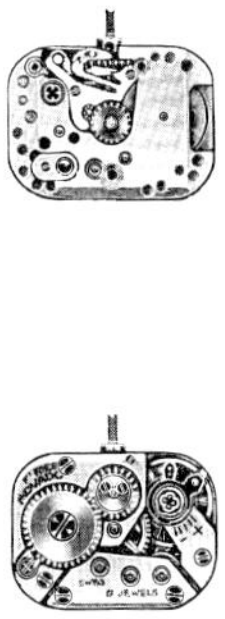

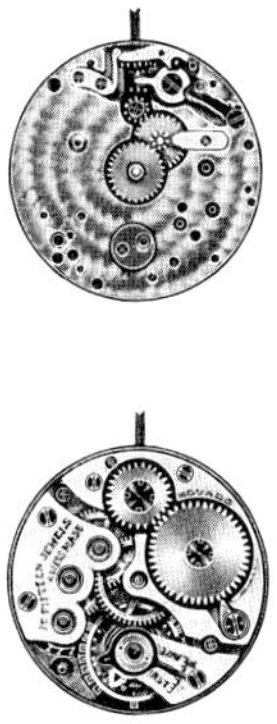

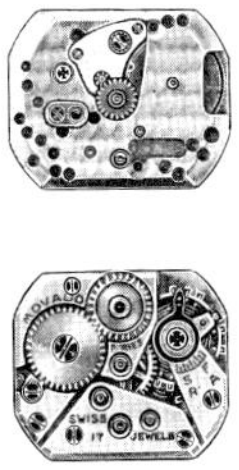

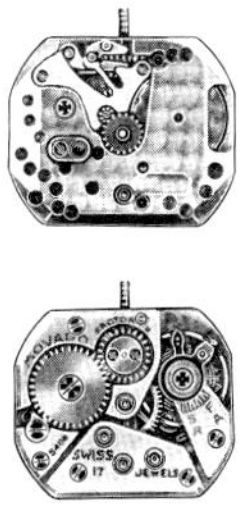

Cal. 65	**Cal. 85**	**Cal. 85** Alarm	**Cal. 90M**	**Cal. 90M, 95M**	**Cal. 99** Silhouette
5′′′ x 7′′′	8¾′′′ – 19,70 mm	8¾′′′ – 19,70 mm	12′′′ – 26,60 mm	12′′′ – 26,60 mm	9′′′ – 20,40 mm
11,50 x 16,00 mm					
	 	 	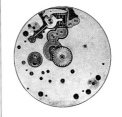 	 	 

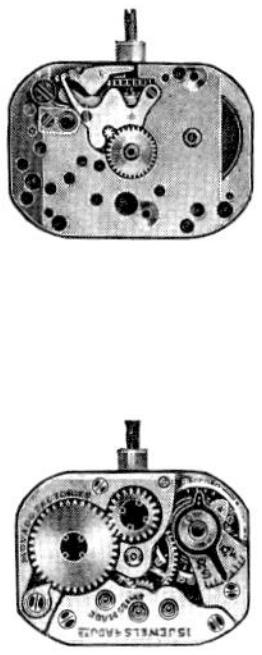

Cal. 105 Ermetolux 8¾′′′ – 19,50 mm	**Cal. 105** **Cal. 107** 8¾′′′ 19,50 mm	**Cal. 115** Futuramic 11½′′′ – 26,50 mm	**Cal. 116** 11½′′′ – 26,50 mm	**Cal. 118** Calendolux 11½′′′ – 26,50 mm	**Cal. 122, 128** Calendoplan 12½′′′ – 28,00 mm
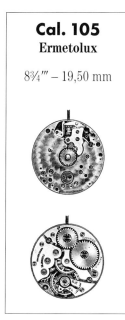	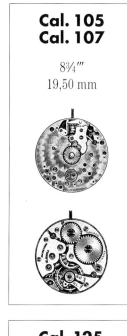	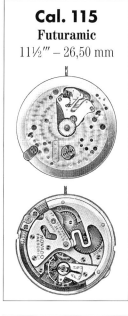	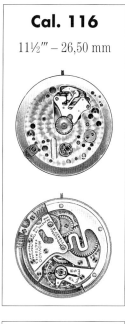	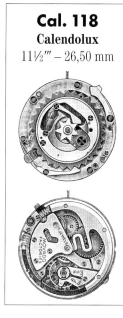	

Cal. 123, 128 S. C 12½′′′ – 28,00 mm	**Cal. 125** 12½′′′ – 28,00 mm	**Cal. 126** 12½′′′ – 28,00 mm	**Cal. 127** 12½′′′ – 28,00 mm	**Cal. 129** Polygraf 12½′′′ – 28,00 mm	**Cal. 135** 12½′′′ – 28,00 mm
	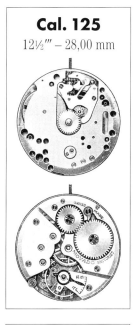	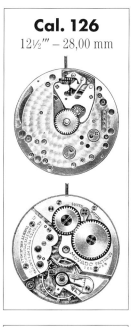	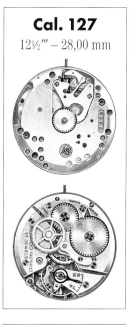	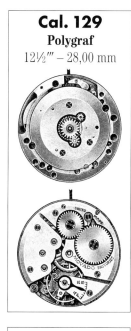	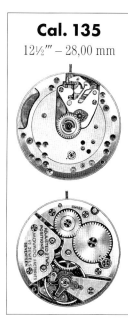

Cal. 150MN **Cal. 157** 10¼′′′ – 22,90 mm	**Cal. 155** Calendermeto 10½′′′ 22,90 x 24,00 mm	**Cal. 159** Ermetoscope 10¼′′′ – 22,90 mm	**Cal. 166** **Cal. 165** Rotor **Automatic Miniature** 7¼′′′ – 16,50 mm	**Cal. 166, 165** **Cal. 169,** **167** Rotor **Automatic Miniature** 7¼′′′ – 16,50 mm	**Cal. 190** 8¾′′′ – 19,50 mm
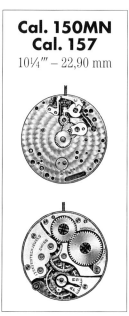	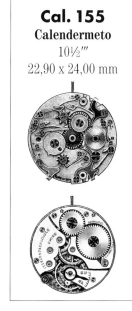			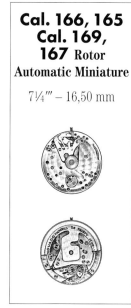	

Cal. 205	**Cal. 220, 220M** **Tempomatic** 220: 12''' – 27,00 mm 220M: 12¾''' – 28,80 mm	**Cal. 221, 226** **Tempomatic** 221: 12''' – 27,00 mm 226: 12¾''' – 28,80 mm	**Cal. 223, 228** **Calendomatic** 223: 12''' – 27,00 mm 228: 12¾''' – 28,80 mm	**Cal. 224, 224A** **222** **Calendoplan Automatic** 12''' – 27,00 mm	**Cal. 225, 225M** **Calendomatic** 225: 12''' – 27,00 mm 225M: 12¾''' – 28,80 mm
10''' – 22,80 mm					

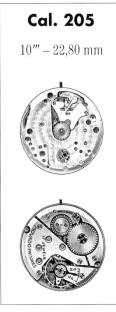 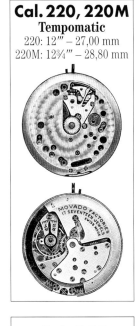 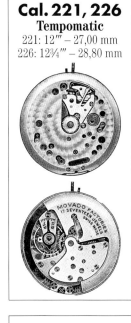   

Cal. 230 **Gentleman** **Automatic** 12½''' – 28,00 mm	**Cal. 245** 8¾''' – 19,40 mm	**Cal. 260M** **Cal. 261, 261A** 10½''' – 23,30 mm	**Cal. 280** 10½''' – 23,30 mm	**Cal. 290** 11½''' – 25,60 mm	**Cal. 298** 11½''' – 25,60 mm

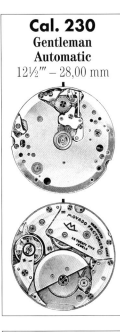 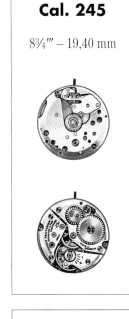 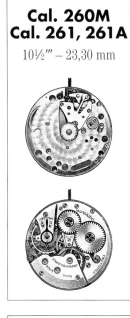   

Cal. 335, 1005 10½''' – 23,30 mm	**Cal. 336** 10½''' – 23,30 mm	**Cal. 345, 1105** 11½''' – 25,60 mm	**Cal. 346** 11½''' – 25,60 mm	**Cal. 348** 11½''' – 25,60 mm	**Cal. 365** 11¼''' – 25,50 mm

  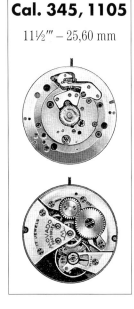   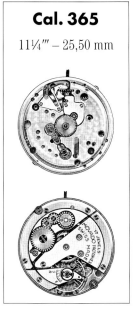

Cal. 368 11¼''' – 25,50 mm	**Cal. 369** 11¼''' – 25,50 mm	**Cal. 375** **Cal. 377** 8¼''' x 9¾''' 18,50 x 22,10 mm	**Cal. 380** Kingmatic «S» 11¼''' – 25,30 mm	**Cal. 388/389** Kingmatic «S» Calendar 11¼''' – 25,30 mm	**Cal. 395** Kingmatic «S» 11¼''' – 25,30 mm
	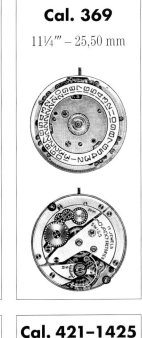		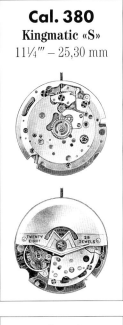	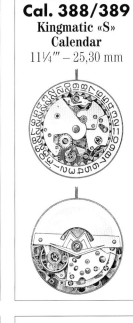	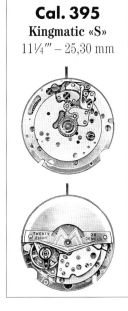

Cal. 400 11¼''' – 25,30 mm 25,50 mm	**Cal. 421–1425** **Cal. 423–1427** Queenmatic 7¼''' – 16,20 mm	**Cal. 425** Queenmatic 7¼''' – 16,20 mm	**Cal. 427** Queenmatic 7¼''' – 16,20 mm	**Cal. 431A** Rotor Automatic 11¼''' – 25,50 mm	**Cal. 431A, 436** Rotor Automatic 11¼''' – 25,50 mm
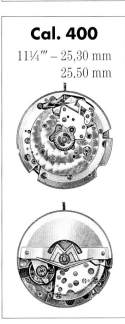	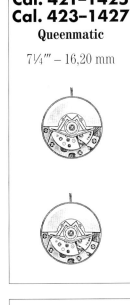	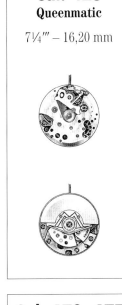	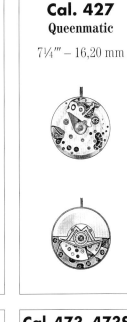	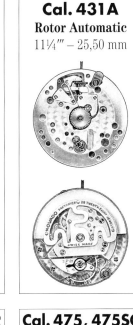	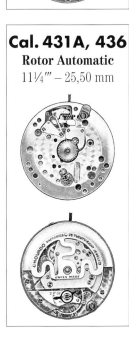

Cal. 438 Rotor Automatic Calendoplan 11¼''' – 25,50 mm	**Cal. 440** **Cal. 443** 6¾''' x 12''' 15,40 x 27,00 mm	**Cal. 470, 477** 10¼''' – 23,10 mm	**Cal. 473, 473SC** Celestograf 10½''' 23,10 x 24,00 mm	**Cal. 475, 475SC** Calendograf 10¼''' x 12½''' 23,10 x 28,00 mm	**Cal. 510** 9''' x 12''' 20,00 x 27,00 mm
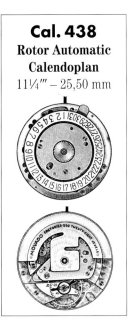	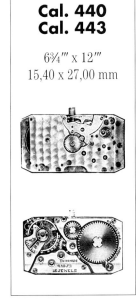	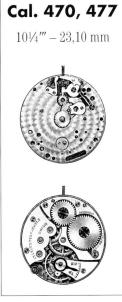	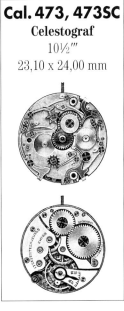		

Cal. 531
Kingmatic
11¾''' – 26,50 mm

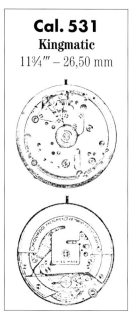

Cal. 531, 536
Kingmatic
11¾''' – 26,50 mm

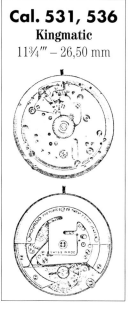

Cal. 538
**Kingmatic
Calendoplan**
11¾''' – 26,50 mm

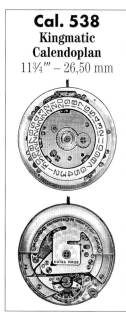

Cal. 540
17''' – 38,00 mm

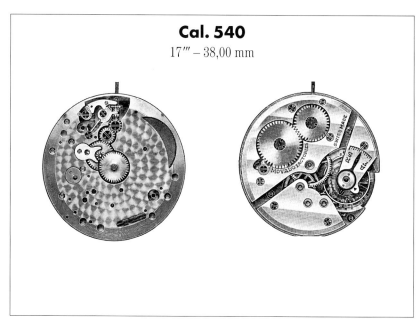

Cal. 545
17''' – 38,00 mm

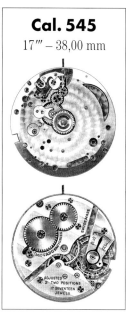

Cal. 575

5¾''' x 7¼'''
13,20 x 16,60 mm

Cal. 575
Ermeto – Baby

5¾''' x 7¼'''
13,20 x 16,60 mm

Cal. 578
Ermeto Calendine

5¾''' x 7¼'''
13,20 x 16,60 mm

Cal. 579
Calendoplan Baby

5¾''' x 7¼'''
13,20 x 16,60 mm

Cal. 608
Tempo-Matic
11½''' – 25,60 mm

Cal. 608
Tempo-Matic
11½''' – 25,60 mm

Cal. 800M
19''' – 42,40 mm

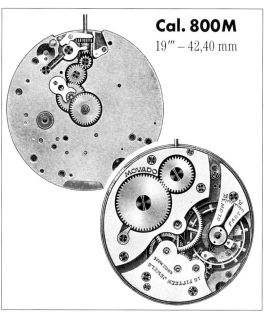

Cal. 895
19''' – 42,30 mm

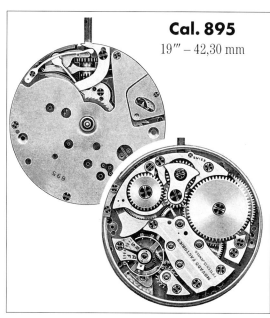

Cal. 900
Ermetophone

11½''' – 25,60 mm

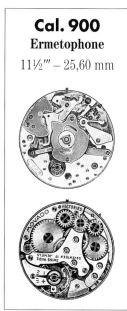

From L'Illustration, Paris 1931

Results of Movado pocket resp. deck watches at Neuchâtel and Kew/Teddington Observatories

The following 3 tables were kindly placed at our disposal by Herr Neumüllers

Key to abbreviations:
P (pocket)   Pocket chronometer
B (bord or deck)   Deck watch

e.g. P 21.51 18: The pocket chronometer was awarded 21.5 points which resulted in 18th place in the first prize group.

264
Movado Observatory pocket chronometer from the 2nd Observatory Series with electrical seconds contacts, made in 1922. Heavy silver case, silvered dial with small seconds, signed "Movado Chronomètre de Bord". Gilded 22''', movement with lever escapement, signed "Movado Factories", movement No. 360 446. Screwed chaton for the escape wheel pivot, aperture in the train wheel cock to observe the seconds contacts. Large Guillaume balance, flat balance spring with double Phillips terminal curves, simple regulator index, 21 jewels, 8 adjustments. This chronometer was awarded a first prize at Neuchâtel Observatory in 1922 and 1925 plus 8th place at Kew in 1923/24.

265
Movement of the deck watch shown in Illus. 122 from the 2nd Observatory Series. Nickel-plated 22''', movement with lever escapement, "filets" decoration, signed "Movado Fy" and "Chronomètre de Bord", movement No. 360 463.

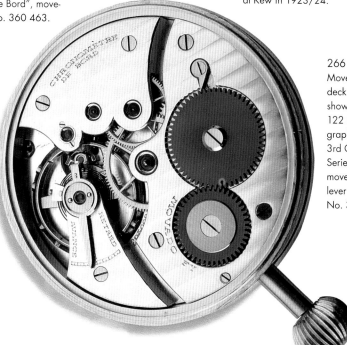

266
Movement of the deck watch shown in Illus. 122 (upper photograph) from the 3rd Observatory Series. Gilded movement with lever escapement, No. 360 516.

MOVADO/Observatory watches   Neuchâtel/KEW

1st Series: 360001 – 45 / (43 mm) / 1912–1922

 2nd Series

	No	1912	1913	1914/15	1916	1917	1918	1919	1920	1921	1922	KEW (Jahr NC Pl)	No
III	01	III	P 15,1 II 44>>		P 8,5 – 68 B>				8,5 – 66 B				..01
II	02		P 13,6 III 74 >	10,5 – 45 B									02
III	03		P 9,9 –143>>	P 16,3 II 94 >				15,4 II 37 B>				1019/20 93,9 22	03
II	04		P 15,7 II 30 >		15,3 II 45 B								04
III	05		P 17,3 I 20>>		8,9 – 66 B>	17,8 I 17 B>							05
III	06		P 13,9 III 68>>		8,7 – 67 B>		19,8 I 3 B						06
I	07		P 7,7 – 188										07
II	08		P 10,1 – 141 >	P 18,6 I 64									08
II	09							P 22,5 I 15 >	14,6 – 31 B				09
III	10		P 12,4 III 100>>						15,0 – 26 B>	14,1 II 36 B			10
o	11												11
III	12		P 27,5 I 3>>	13,3 II 28 B									12
III	13		P 10,2 – 135>>		P 19,2 I 37 >	18,8 I 13 B							13
III	14	P 13,2 III 65 >	9,5 – 48 B>					16,9 II 29 B					14
II	15	P 10,0 – 133 >	P 8,8 – 167									? 1929 94,8 13	15
II	16	P 10,4 – 199 >	P 13,1 III 88										16
IIII	17	P 7,1 –186>>>	P 10,8 –127>>	P 17,2 I 86 >					9,8 – 60 B				17
II	18		P 12,8 III 94 >	18,7 I 10 B									18
II	19	P 8,7 – 161 >			P 13,1 III 85								19
III	20		P 10,1 – 136>>					P 16,6 II 40 >	15,1 – 25 B>			1919/20 92,3 36	20
II	21	P 12,5 III 80 >	15,2 II 20 B										21
II	22	P 14,4 II 48 >	11,7 III 36 B									1919/20 92,5 21	22
III	23					P 11,0 – 99>>		P 14,2 III 59 >	16,5 II 7 B>				23
IIII	24	P 6,4 –196>>>	P 7,8 –184>>		P 22,7 I 15 >	14,0 II 37 B							24
II	25		P 11,0 – 122 >		P 18,1 I 43								25
III	26				P 17,7 I 47 >>			12,1 – 61 B>	14,9 – 27 B				26
III	27		P 10,8 – 126>>				P 21,1 I 24 >	21,3 I 11 B>				1918/19 94,8 7	27
II	28		P 14,9 II 53 >					10,4 – 75 B>				1919/20 93,1 31	28
I	29		P 14,6 II 58										29
I	30		P 10,4 – 131										30
II	31						P 26,5 I 4 >	16,2 II 32 B>				1918/19* 94,4 10	31
II	32				P 11,1 –110 >				P 17,6 – 73				32
II	33				P 19,6 I 32 >			17,5 I 24 B>				1919/20 90,6 51	33
I	34								P 21,8 I 28				34
III	35							P 15,9 II 46>>	15,3 – 21 B>		12,8 – 54 B		35
III	36			P 17,8 I 73>>	14,4 II 50 B>						P 10,8 – 87		36
II	37					P 26,2 I 4 >		13,9 III 50 B>				1919/20 93,7 24	37
II	38						P 19,8 I 28 >	17,2 I 27 B					38
I	39			P 25,7 I 15									39
I	40								P 20,1 I 35				40
II	41					P 20,3 I 22 >	15,8 II 12 B>					1918/19 94,8 6	41
II	42			P 17,7 I 76 >				13,0 III 55 B>				1919/20 94,3 17	42
II	43			P 21,8 I 31 >		P 8,8 – 118							43
III	44				P 11,8 –104 >>	P 18,8 I 33 >		19,5 I 14 B					44
o	45												45
		8	22 + 1* 3	7 3	8 5	5 3	3 + 1* 2 4	11	3 8	1	1 1		
			*322962 P 8,8 – 168				*360433 (s.a. 2. Serie) P 27,3 I 2 >						
Regl.		Edmond Ditesheim, Ernest Frey	E.D. / E.F.	E.D. / Charles Haenggeli	E.D. / Ch. H.	E.D. / Louis Augsburger	E.D. / L.A.	E.D. / L.A.	E.D. / L.A.	E.D. / L.A.	E.D. / L.A.		

MOVADO/Observatory watches

2nd Series: 360434 – 478 (50 mm) / 1919–1927

Neuchâtel/KEW

Rule change
Diameter information

3rd Series

M* = Chronometre Marine

	No	1919 Poche (NC Pr Pl) / Bord	1920 Poche / Bord	1921 Poche / Bord	1922 Poche / Bord	1923 Poche / Bord	1924 Poche / Bord	1925 Poche / Bord	1926 Poche / Bord	1927 Poche / Bord	KEW Jahr	NC	Pl	No
II	360434	P 29,2 I 3 >		27,9 I 2 B >							1922/23	92,6	28	..34
III	..35		P 24,4 I 19>>	21,4 I 14 B >	11,0 – 72 M* >						19207/21	94,5	4	35
IIII	36	P 17,3 II 36>>>	14,1 – 3 B >	21,0 I 18B >	7,2 –116 M* >						1921/22	95,6	5	36
II	37	P 13,7 III 64 >	23,8 I 2B>								1919/20	92,8	33	37
III	38		P 21,4 – 61 >>	P 24,8 I 8 >	15,9 II 40 B									38
III	39		P 14,5 – 91 >>	P 17,5 I 31 >				11,3 II 72 B						39
II	40		P 22,6 I 24>	16,5 II 32B >							1921/22	93,2	36	40
II	41		P 27,0 I 9>	20,8 I 19 B >							1921/22	95,2	12	41
III	42		P 11,5 –116 >>	P 21,1 I 19 >					7,4 I 12B					42
III	43		P 10,2 –132>>					P 4,8 I 9 >	7,7 I 22B		1923	95,2	13	43
II	44			P 15,1 II 45 >	14,9 II 45 B >>						1924/25	94,7	5	44
III	45		P 22,7 I 23>>	18,0 I 28 B >	8,1 – 107 M*						1921/22	92,9	38	45
III	46			P 23,6 I 13 >>	11,7 (14,6) –107 B >			9,5 I 38 B >			1923/24	95,5	8	46
II	47				P18,1(7,7) I 35 >				7,6 I 19B					47
II	48			P 12,5 III 65 >			7,8 I 18 B >				1923/24	90,1	47	48
II	49			P 16,3 II 38 >			7,6 I 15 B >				1924/25	93,9	13	49
0	50													50
I	51									(1930: P 9,6 II 37)				51
II	52			P 19,5 I 23 >				7,9 I 15 B						52
I	53					P24,2 (5,5) I 12 >					1922/23	95,5	6	53
0	54													54
II	55			P 25,9 I 10 >					7,5 I 17B					55
II	56			P 21,5 I 18 >		23,4 (6,6) I 4 B								56
II	57				P27,5 (5,2) I 10 >			9,9 II 44B						57
I	58				P 18,1 I 29									58
II	59				P 19,3 I 24 >		7,5 I 12 B >				1923/24	94,1	28	59
II	60				P 28,1 I 8 >			6,5 I 5 B						60
II	61				P 17,2 I 32 >			8,3 I 20 B						61
II	62				P 16,6 II 37 >	19,3 (7,5) I 7 B >					1923/24	94,8	18	62
I	63							P 4,9 I 11						63
II	64							P 5,3 I 15 >		10,0 II 28 B				64
II	65					P22,6 (5,8) I 14	8,7 I 12 B >							65
I	66								P 5,3 I 14					66
II	67						P 5,4 I 17 >	7,8 I 14 B						67
II	68						P 5,0 I 17 >	6,0 I 3 B >			1924/25	94,2	9	68
II	69							P 6,1 I 29 >	8,1 I 14B					69
0	70													70
I	71				P 23,0 (6,5) I 22									71
0	72													72
I	73							P 5,4 I 17						73
I	74								P 6,4 I 34					74
I	75						P 5,8 I 23							75
I	76									(1928: P 8,3 I 38)				76
I	77						P 5,6 I 20							77
II	78					P19,2 (7,4) I 32 >	7,1 I 10 B >				1924/25	94,6	7	78
II	*360433/Tourb.	14,9 II 44 B >									1919/20	95,9	3	360433
III	*26217/ressort/50 mm	P 13,7 III 66 >>					7,7 II 78 M* >				1919/20	94,0	21	26217
II	*6664/ressort/50 mm		18,1 I 3 B >					P 5,1 I 13 >	4,7 I 1B >		1927	95,5	5	6664
		3 + 1* (1*)	8 2+1*	3 6	12 2+3* *M*	6 3	4 5+1* *M*	4+1* 7	3 5+1*	– 2				
		Edmond Ditesheim / Louis Augsburger	E.D. / L.A.	E.D. / L.A.	E.D. / L.A.	E.D. / L.A.	E.D. / L.A.	E.D. / L.A.	E.D. / L.A.	E.D. / L.A.				

MOVADO/Observatory watches — Neuchâtel/KEW

3rd Series: 355001 – 10/ 360481 –519 / (64/65 mm) / 1927 –1939

	No	1927 Poche / Bord	1928 Poche / Bord	1929 Poche / Bord	1930 Poche / Bord	1931 Poche / Bord	1936 Poche / Bord	1938 Poche / Bord	1939 Poche / Bord	KEW Jahr	KEW NC	KEW Pl	KEW No
II	355001	P 6,6 I 28 >		6,4 I 6B>						1930	92,2	22	..001
I	..002								P 7,9 I 5				002
II	003	P 5,9 I 18 >	6,6 I 5 B										003
II	004			P 4,1 I 5 >	5,5 I3B>					1930	95,7	7	004
I	005							P 5,3 I 4					005
III	006	P 4,3 I 4>>	8,8 I22 B>	(>)				6,6 – 4 B		1930	96,5	1	006
II	007	P 5,8 I 16 >	7,7 I11 B>							1929	90,4	22	007
II	008	P 5,3 I 10 >	7,3 I 8 B>							1928	96,0	1	008
II	009		P 5,5 I15 >		6,3 I 8 B>					1931	95,2	12	009
o	010												010
II	360481			P 5,5 I 15 >		6,1 I 4 B							...481
I	...482							8,7 III 3 B					482
I	483			P 6,6 I 25									483
II	484	P 4,8 I 7 >	7,4 I 9 B>							1929	86,1	27	484
o	485												485
II	486	P 5,4 I 11 >		6,9 I 12 B									486
II	487	P 5,8 I 14 >	6,4 I 4 B>							1929	96,4	3	487
II	488		P 4,2 I 8 >>	6,9 I 11B>						1930	96,1	4	488
I	489			P 4,0 I 4									489
II	490			P 6,0 I 20 >			4,2 I 1 B						490
II	491		P 4,6 I11 >		6,4 I 9 B								491
II	492	P 5,5 I 12 >				6,1 I 5 B							492
II	493					P 4,6 I 12 >	7,1 II 4 B						493
III	494		P 6,0 I 19 >>		7,5 I14 B>>			P 8,5 – 27					494
II	495			P 4,5 I 8 >	5,7 I 4 B>					1930	94,6	13	495
II	496				P 5,0 I14 >			7,1 II 2 B		1931	94,7	18	496
II	497				P 4,8 I10 >			P 9,3 – 28					497
o	498												498
I	499					P 3,8 I 8 >				1931	95,4	9	499
I	500				P 4,8 I12								500
I	501						P 4,1 I 1						501
II	502				P 4,7 I 9 >	6,9 I 11 B				1931	93,7	28	502
I	503					P 3,3 I 2 >		P 7,3 – 26					503
II	504						P 5,6 I 7 >						504
o	505												505
o	506												506
o	507												507
o	508												508
o	509												509
o	510												510
o	511												511
o	512												512
I	513						P 6,1 I 10 >	P 6,5 I 9					513
II	514						P 5,0 I 2	P 6,2 – 25					514
I	515												515
o	516												516
I	517							P 5,7 I 7					517
o	518												518
I	519						P 5,1 I 3						519
		+(2*)	4+1* 6	6 3	4+1* 5	3 3	5 2	7 3	1 –				
		*360464/69/ 50mm	*360476/50mm		*360451/50mm								
		Edmond Ditesheim Louis Augsburger	E.D. L.A.	E.D. L.A.	L.A.	L A	Werner-A. Dubois	W.-A.D.	W.-A.D.				

Movado Observatory wrist chronometers and the years of their participation in the Chronometer Trials at the Observatories of Neuchâtel and Geneva

1. The three-digit number series with the round Ø 30 mm calibre.

In brackets: Geneva Observatory, without brackets: Neuchâtel Observatory.

Movement No.	Participation year
201	1955, 1958
202	1957
203	1957
204	1956, 1957, 1958, 1959
205	1957
206	1955, 1962, 1964
207	1956, 1962, 1963
208	1958, 1959
209	1956, 1957, 1958
210	1956
211	1963, 1964
212	1956, 1957
215	1962, (1962), 1963
216	1957, 1959, 1963
217	1963
218	1961, 1962
219	1958
220	1958
221	1958, 1962, (1962)
222	1958
223	1959
224	1958, 1959, 1963
225	1963, 1964
226	1961, 1963
227	1962, (1962)
228	1962, 1963
229	1961, 1962
230	1962, (1962)
231	1961, (1962), 1963
232	1963
234	1961
236	1961, 1962, 1963
237	(1962), 1963

2. The six-digit number series with the rectangular calibre.

In brackets: Geneva Observatory, without brackets: Neuchâtel Observatory.

Movement No.	Participation year
270 014	1964
270 016	1966
270 018	1964, 1965
270 019	1964, 1965, 1966
270 020	1964, 1965
270 021	1964, 1965, 1966
270 022	1965, 1966
270 023	1964, 1966
270 025	1964, 1965, 1966
270 026	1964
270 027	1964, 1966
270 029	1966, (1966)
270 030	1965, 1966
270 032	1965
270 033	1965
270 034	1965, 1966, (1966)
270 035	1965
270 037	1965
270 038	1965
270 039	1965, 1966
270 042	1966
270 043	1966

Trial results of Movado wrist chronometers at Neuchâtel Observatory

267
Skeletonised Observatory
movement, Calibre 30 P
(Peseux 260), owned by
Henri Guye (adjuster).

1950
Not placed: Nos.
330 003, 330 004, 330 007

1956
Series prize for the four
best wrist chronometers.
First prizes:
1st place for	No. 210
Third prizes:	
50th place for	No. 206
51st	No. 207
Not placed: Nos. 209, 212	

1957
Series prize for the four
best wrist chromometers.
1st place for	No. 202
2nd	212
3rd	205
5th	203
9th	209
16th	216
19th	204

1958
1st prizes:
1st place for	No. 222
3rd	219
6th	209
11th	220
19th	208
21st	221
24th	224
39th	201
Not placed:	No. 204

1959
1st prizes:
2nd place for	No. 204
6th	208
21st	223
3rd prizes:	
99th place for	No. 224
Not placed	No. 216

1961
1st prizes:
21st place for	No. 231
46th	226
57th	236
59th	234
67th	229
82nd	218

1962
1st prizes:
10th place for	No. 227
11th	230
38th	207
40th	221
43rd	215
49th	236
53rd	228
72nd	206
83rd	229
87th 218	

1963
Placed:
1st place for	No. 226
56th	228
57th	216
64th	224
68th	225
71st	232
80th	211
86th	215
97th	217
117th	231
123rd	236
Not placed: Nos. 207, 237	

1964
Placed:
23rd place for	No. 270 020
58th	270 019
60th	270 025
63rd	270 021
73rd	206
74th	270 023
76th	211
88th	270 026
98th	270 018
118th	270 014
Not placed: Nos. 225, 270027	

1965
Placed:
56th place for	No. 270 038
61st	270 019
67th	270 034
71st	270 021
84th	270 022
94th	270 037
97th	270 033
102nd	270 018
103rd	270 030
113th	270 035
122nd	270 032
Not placed: Nos. 270 020, 270 039, 270 025	

1966
Placed:
21st place for	No. 270 029
37th	270 034
67th	270 042
91st	270 022
103rd	270 016
117th	270 027
140th	270 039
157th	270 043
Not placed: Nos. 270 019, 270 021, 270 030, 270 025, 270 023	

Movado inventions and patents
(Swiss patents only)

Patent No. 27 690, dated 17th January 1903

This patent concerns a space-saving simple date indication mechanism that may be mounted on a plate which may in turn be fitted to the pillar plate of a pocket watch movement. The plate may be fixed by two screws into a recess turned in the pillar plate, beneath the dial, and may be combined with a repeating movement (not shown) at the same level, using less than half the plate area. This method facilitates the manufacture of thin pocket watches, which were currently fashionable, notwithstanding the repeating and date indication.

Patent No. 28 678, dated 9th June 1903

This patent concerns a winding and hand-setting mechanism for pocket watches.

It is a variant of the clutch system of winding and hand setting whereby the winding crown (respectively the winding stem) is either pulled out or pushed in. In Switzerland this principle became the final solution after 1900 and was later also adopted for wrist watches. The novelty of this construction by Achille Ditesheim is that it may be applied to the European form of one-piece winding stem (Fig. 3) or the American pattern with a divided winding stem (Figs. 1-2). In the case of the European version the setting lever spring, S, is simply not required. This spring ensures the position of the setting lever, A, and the stem, T, with the divided winding-stem system.

Patent No. 28 679, dated 9th June 1903

This patent is for an improved method of fitting a dial without pillars, or feet, to a watch movement; a so-called "rim fitting dial".

The applicant, Achille Ditesheim, explains that employing the usual solution with feet on the back of the dial which project into the watch plate and are then secured therein has the disadvantage that there may be difficulty centring the dial, which frequently leads to rejected parts.

He proposes a dial without feet which is snugly fitted into an exactly made metal ring. This assembly is then fitted to the front plate. Contrary to other such dials which are sprung onto the plate, this type is set on an emplacement and secured laterally by means of a pin and a screw. This expensive construction is only applicable to valuable pocket watches but offers the best possible protection to delicate enamel dials.

Patent No. 37 776, dated 29th October 1906

This patent concerns a special form of fine adjustment regulator index for precision watches. It is based on the principle of a worm, or endless screw (i), which engages a toothed segment (g) of the ring (d) in such a way as to turn the latter. A gap on the opposite side of the ring embraces the regulator index, which is thus caused to move with the ring. The semi-circular spring (l) applying pressure to the index is not strictly necessary with this construction and may be dispensed with, a point made in the patent, since the index is securely guided in both directions by the toothed segment. There are many such precision regulator index sytems and, in certain circumstances, a patented construction of one's own can save licensing fees.

Patent No. 45 161, dated 13th November 1908

This patent is the first of a series of five pocket watch case constructions which aim to give a flat case as well as economising on material.

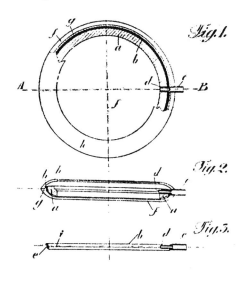

This first patent shows a two-part case whereby the movement is fitted to a solid ring of non-precious metal. The ring is surrounded by a narrow precious metal ring with two snaps. This movement ring is pressed into the case back and then the bezel can be snapped on. This offers a simple, flat construction, without a hinge. However, it is further explained that this case construction can also have a hinged hunter cover over the bezel and glass.

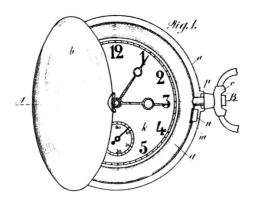

Patent No. 45 162, dated 21st November 1908

This second watch case patent continues with two case halves which enclose a movement ring and is particularly concerned with the connection of the pendant, which is soldered to the movement ring, which is in turn fitted to the case. The pendant is surrounded by a special form of collar which is normally known as "à goutte", which forms a part of the case halves.

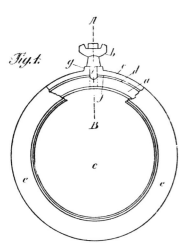

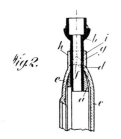

Patent No. 45 667, dated 19th December 1908

This patent concerns a special form of double cover hunter case for pocket watches. Unusually the movement is not accessible from the back but only from the front in the manner that is common in old English watches and some American watches. The movement and dial are once again fixed to a stable movement ring (d). This assembly is then fitted into a suitable hollow in the case back where it is retained by snapping on a likewise stout bezel (g). After removing the bezel the movement capsule may be taken out of the case. This is a construction in which the advantages over the normal hunter case, where the dial is accessible via the bezel and the movement from the back, are not immediately obvious other than, perhaps, that there is a saving of one cover, thus making the case

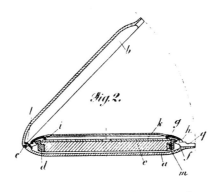

flatter. The usual elegance of the hunter case is somewhat lost due to heaviness of the bezel, which not only carries the glass but must also be sufficiently stable to retain the entire movement.

Patent No. 47 357, dated 29th May 1909

Once again this patent concerns a pocket watch case in the form of a double cover hunter case. Of the two covers, the one over the dial has a circular glazed opening to read the time. The size of this opening lies somewhere between

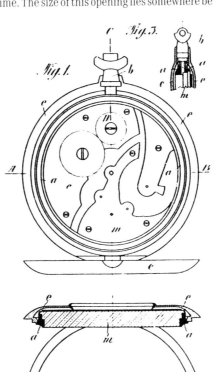

that of the usual opening in a half hunter case and that of a normal bezel. Movado exercised considerable fantasy in the various forms of this glazed opening and therefore likewise the dial. The movement is once more fitted in a ring placed between the two covers but may be observed or removed from the back of the case.

Patent No. 51 670, dated 27th April 1910

This is the fifth patent for a type of hunter pocket watch case. As with Patent No. 45 667, the ring with movement and dial form an independent capsule. After opening the sprung front cover, this capsule may be removed from the front of the case. However, in this instance the bezel is also fitted to the movement capsule and the complete

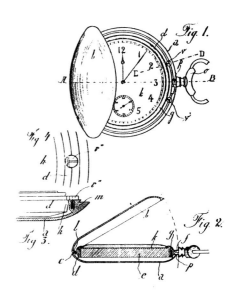

unit is secured to the case back by two screws, 'v v'. The screws are so shaped that a part turn releases the movement capsule.

Patent No. 60 360, dated 7th June 1912

This patent with the blunt title "Montre-bracelet" (wrist watch) concerns the construction of the "Polyplan". We judge it to be so significant that it warrants a complete translation:

"There are already existing wrist watches with elongated rectangular, curved cases, enclosing the movement, however, they do not conform sufficiently well to the contours of the arm. Until now the movements of these watches have been round, that is to say they have been constructed on a large plate such that only a small portion of the available space in the case may be utilised.

With the aim of using in an optimal manner the available space of the case, which is of an elongated form and curved to suit the shape of the human arm, the movement of the wrist watch is mounted on a plate which is angled in such a way as to form several adjacent faces placed end to end, like a prism. For example, the plate could be presented as having two angled ends and a middle portion which carries all the mobiles of the movement, except that of the balance wheel. The two inclined end portions could carry the majority of the parts associated with the winding and hand setting at one end, and the other end could carry the balance staff and its accessories. The lever of the escapement could be cranked to compensate for the difference in level between the escape wheel and the organs corresponding to the balance staff.

The attached drawings, figs. 1 and 2, for example, show a side view and a plan view of the object of this invention as applied to an arm, the latter shown in section by fig. 1.

Fig. 3 shows a plan view of the movement of the watch and fig. 4 a longitudinal section of the case enclosing the said movement.

Thus it may be seen from these figures that the wrist watch is composed of an elongated case A, furnished with stirrups, or lugs, at the opposing ends 9 which are connected by means of a flexible strap 10 with a buckle 11 and a sliding loop 12. The lower face of this case, that is to say the face on the opposite side of the dial a, is curved to conform to the arm whilst the said dial, likewise the hands, b c, which move in front of the former, are protected by an elongated glass d.

The large plate, B, of the movement is formed by a bent strip, that is to say obtusely angled in two places. The centre portion c, set between the two ends i j, angled relative to the former, carries a small, parallel, pentagonal plate J. The axes of the centre wheel D on the centre pinion f, which engages the great wheel E on the barrel C, are placed between B and J. The barrel arbor g of the barrel C carries the winding wheel G below the large plate B. Winding wheel G engages an intermediate wheel H, secured below the end portion i of the plate B and parallel to the said portion i, by a round plate y secured by two screws. Wheel H normally engages winding pinion I fitted on the winding stem L. The faces of the toothed wheels G and H consequently form an obtuse angle to one another. The winding click k, together which its actuating spring l, is fitted beneath the centre portion e of the plate B.

Above the small plate J the centre arbor, carrying centre wheel D, also carries a pinion m, linked by the two intermediate wheels n o, maintained parallel, on J to the portion e of plate B, to a wheel K which is solid with a pinion p and maintained parallel with the latter, to the extremity i of the plate on a post M which is perpendicular to the surface of i.

The winding stem L is guided partly in the post M and partly in a bearing support N formed in portion i of the plate B. The stem terminates with a winding crown 15 placed in the cavity O of the case A. Fitted on a square portion of the stem L is a double sliding Breguet pinion P, which can engage alternately with the winding pinion I or with the pinion p according to whether the stem L is in the position for winding or set hands. Normally the sliding pinion P engages with the winding pinion I, as is shown in the drawing. If one wishes to place it in the position to set the hands it is only necessary to pull the stem in the direction of the arrow x so as to cause the annular groove q to act on the rocking piece r, which pivots on portion i of B, and is acted upon by a spring s and actuated by an arm t on a rocking piece u pivoted on portion i of B under the action of a spring w fitted to the same portion i of B and penetrating the annular groove v of sliding pinion P. When the hand-setting operation has been completed the sliding pinion P may be returned to the winding position by pushing the stem L in the opposite direction to the arrow x.

One may see from the preceding description that the major part of the components serving the winding- and hand-setting operations are carried by the end portion i of the plate B.

As the watch runs, the centre wheel motion is transmitted via the third pinion and wheel, 1 and 2, to the fourth pinion and wheel, 3 and 4, and then to the escape pinion 5 and the escape wheel 6. The axes of these three series of wheels and pinions are placed between the cen-

tre portion e of the plate B and a second small plate U, of generally triangular form with a lateral cut-out 16, fitted to the plate B and parallel to the latter. The pallets V, operating in conjunction with the escape wheel 6, are fitted between the centre portion e of the plate B and a small curved bridge 7 screwed to the plate, and engage with the components of the balance staff 14 by means of a cranked lever and fork 8. The balance staff 14 is fitted between the inclined end j of the plate B and a cock y, screwed to the said portion j, and parallel to the latter. The balance wheel is combined in the ordinary way with the usual accessories such as balance spring, regulating index, etc.

One may see from the preceding description that the balance staff is assembled on the portion j of the plate B at an acute angle to the axes of the other mobiles of the movement on the centre portion e of the said plate.

One may see that with this arrangement almost all of the interior space of the case is utilised to contain the movement and that a case of the dimensions given could, for example, receive a normal 8‴ movement with a circular plate whereas the components of the movement described are comparable with those of a normal movement with an 11‴ circular plate. It follows that the watch with a case containing a movement as described will have a more precise rate as well as continue to function for an extended period of time. The wrist watch described may be constructed with or without motion work. If it is made with motion work the latter may be placed, as usual, between the main plate and the dial as shown in fig. 4.

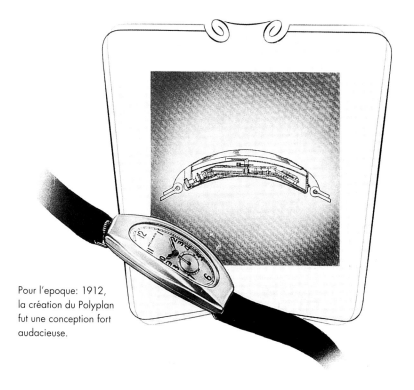

Pour l'epoque: 1912, la création du Polyplan fut une conception fort audacieuse.

List of patents granted to L.A. & I. Ditesheim and Movado

Patent No.	Dated	Original title	English title and remarks
9 163	25. 10. 1894	Nouveau système d'encliquetage pour montres de tous calibres	New barrel ratchet system for movements of all calibres
18 863	11. 2. 1899	Perfectionnement aux machines à graver et à guillocher	Improvements to engraving and engine-turning machines
26 961	22. 9. 1902	Machine à faire le décor dit mille feuilles	Machine to produce "mille feuilles" decoration
27 690	17. 1. 1903	Montre à répétition avec mécanisme de quantième simple	Repeating watch with simple date mechanism (see Illus. 23)
28 678	9. 6. 1903	Mécanisme de remontoir et de mise à l'heure	Winding and hand-setting mechanism (see Illus. 24)
28 679	9. 6. 1903	Perfectionnement aux mouvements de montres	Rim-fitting dial
28 804	17. 7. 1903	Perfectionnement au mécanisme de mise à l'heure de montres remontoirs savonnettes	Improvements to hand-setting mechanism for hunter cased watches
30 608	2. 5. 1904	Montre avec raquetterie perfectionnée	Pocket watch with externally operated regulator index
34 976	1. 12. 1905	Coq de montre	Attachment of balance spring stud to the cock
37 776	29. 10. 1906	Raquetterie	Fine adjustment regulator index (see Illus. 22)
41 143	24. 9. 1907	Boîte de montre	Watch case with eccentric dial and precious stones
42 092	12. 2. 1908	Bracelet	Specially decorated wrist watch bracelet
44 562	12. 10. 1908	Montre bracelet	Strap attachment for wrist watches
45 161	13. 11. 1908	Boîte de montre	Specially decorated wrist watch bracelet
45 162	21. 11. 1908	Boîte de montre	Special two-part pocket watch case (see Illus. 268)
45 667	19. 12. 1908	Montre avec boîte savonnette	Special hunter watch case (see Illus. 269)
47 020	17. 5. 1909	Bouton de manchette	Cuff links with a watch
47 357	29. 5. 1909	Boîte de montre	Special pocket watch case (see Illus. 270)
51 670	27. 4. 1910	Montre avec boîte savonnette	Special hunter watch case (see Illus. 271)
60 360	7. 6. 1912	Montre-bracelet	Construction of the Polyplan (see Illus. 33)
104 595	22. 5. 1923	Pièce d'horlogerie de précision à ressort moteur	Precision pocket watch with a small eccentric movement
108 966	2. 3. 1925	Montre de poche	Pocket watch case without hinges or pendant
122 390	16. 9. 1927	Montre	Pocket watch version of the "Valentino" (see Illus. 76)
125 036	16. 3. 1928	Montre	Two-part flat watch case
127 820	1. 10. 1928	Pièce d'horlogerie à remontoir	Automatic rack winding of the "Ermeto" (see Illus. 106)
130 940	16. 3. 1929	Montre	Alternative dial mounting for the "Polyplan" (see Illus. 34)
135 524	2. 12. 1929	Montre	Attachment loop for the "Valentino" (see Illus. 75)
137 674	1. 4. 1930	Pièce d'horlogerie à remontoir	Mantel clock with slide winding
137 904	16. 4. 1930	Sacoche	Shoulder bag with integrated Ermeto
138 641	16. 5. 1930	Montre briquet	Table lighter with watch
140 801	1. 9. 1930	Montre-bracelet	Wrist watch version of the "Valentino" (see Illus. 77)
143 864	30. 11. 1930	Nécessaire avec montre	Sewing kit with integrated Ermeto
172 421	15. 10. 1934	Montre-bracelet	Flexible strap lugs for a wrist watch
188 661	15. 1. 1937	Boîte de montre	Back of a wrist watch with interchangeable decoration
191 277	15. 6. 1937	Compteur de temps	Construction of the "Cronoplan" (see Illus. 146)
192 625	31. 8. 1937	Mouvement de montre-chronographe	Chronograph mechanism on its own plate
201 997	31. 12. 1938	Montre	Special wrist watch case for a form movement
211 977	31. 10. 1940	Montre de forme oblongue	Special oblong wrist watch case
214 436	30. 4. 1941	Mouvement de montre ancre à seconde	Wrist watch form movement
215 177	15. 6. 1941	Boîte de montre savonnette	Hunter case for wrist watches
220 535	15. 4. 1942	Pièce d'horlogerie à calendrier automatique	Wrist watch form movement with day and date shown in apertures
220 799	30. 4. 1942	Montre	Two-part watch case
226 490	15. 4. 1943	Mouvement de montre à remontage automatique par masse oscillante	Automatic winding with a swinging weight (see Illus. 180)
247 036	15. 2. 1947	Pièce d'horlogerie à remontage automatique et calendrier	Round wrist watch movement with automatic winding & calendar
256 594	31. 8. 1948	Mouvement d'horlogerie à balancier	Method for protecting a watch balance spring
260 909	15. 4. 1949	Montre-étui avec calendrier	Ermeto with calendar ("Calendermeto")
265 568	15. 12. 1949	Montre	Special plate construction to facilitate checking the third wheel pivots

List of patents granted to L.A. & I. Ditesheim and Movado

(continued)

Patent No.	Dated	Original title	English title and remarks
274 908	30. 4. 1951	Pièce d'horlogerie à calendrier	Date displayed in an aperture
278 789	31. 10. 1951	Dispositif avec montre, pour poste téléphonique	Public telephone with a watch
279 370	30. 11. 1951	Pièce d'horlogerie	Mechanism of the world time watch "Polygraph" (see Illus. 194)
279 956	31. 12. 1951	Pièce d'horlogerie à remontage automatique	Particularly flat wrist watch with automatic winding (Cal. 115 "Futuramic"), (see Illus. 181)
281 490	15. 3. 1952	Pièce d'horlogerie à remontage automatique par masse oscillante	Shock absorbing spring mounting for the winding weight of Cal. 115 "Futuramic" (see Illus. 184)
282 451	30. 4. 1952	Pièce d'horlogerie	Fixing of a wrist watch movement in the case
286 184	15. 10. 1952	Montre-réveil	Ermeto with alarm "Ermetophon"
289 105	28. 2. 1953	Dispositif de fixation, à un bâti de pièce d'horlogerie, d'un cadran muni de pieds	Special system for fixing a dial with feet
298 561	15. 5. 1954	Pièce d'horlogerie à seconde au centre	Centre seconds with a special system for fixing the hand arbor
302 612	31. 10. 1954	Pièce d'horlogerie à remontage automatique par masse rotative	Automatic winding with a central rotor
313 606	30. 4. 1956	Dispositif de fixation d'une masse rotative de remontage de pièce d'horlogerie	Jewelled bearings for the rotor of an automatic winding movement (for Raymond Polo)
319 691	28. 2. 1957	Montre-bracelet	Continuous watch strap for a wrist watch (for Bernard Ditesheim)
322 337	15. 6. 1957	Montre à boîte enfermeé dans un étui protecteur	Stand for the "Ermeto" (for Bernard Ditesheim)
323 047	15. 7. 1957	Montre à boîte enfermeé dans un étui protecteur	Winding mechanism similar to that of the "Ermetophon" (for Bernard Ditesheim)
324 269	15. 9. 1957	Montre-réveil à boîte enfermeé dans un étui protecteur	Winding mechanism for the "Ermetophon" (for Bernard Ditesheim)
324 271	15. 9. 1957	Pièce d'horlogerie à calendrier	Calendar mechanism with the date in an aperture (for Raymond Polo)
335 187	31. 12. 1958	Dispositif d'entretien électromagnétique du mouvement d'un balancier-spiral	Electromagnetic power source for a timepiece with balance and spring
337 785	15. 4. 1959	Pièce d'horlogerie à remontage automatique	Automatic winding with a ball bearing rotor (for R.Polo)
339 140	15. 6. 1959	Dispositif d'entretien électromagnétique du mouvement d'un ensemble balancier-spiral	Electromagnetic power source for a timepiece with balance and spring
347 138	15. 6. 1960	Montre à remontoir automatique	Bearings for an automatic winding rotor (for R.Polo and Marcel Loichat)
351 221	31. 12. 1960	Pièce d'horlogerie	Date in an aperture controlled via the winding crown (for Raymond Polo)
352 288	15. 2. 1961	Dispositif de remontage et de mise à l'heure d'une montre	Winding and hand setting from the back of a watch (for R.Polo)
353 691	15. 4. 1961	Dispositif d'entretien électromagnétique d'un ensemble balancier-spiral	Electromagnetic drive for a balance and spring
355 406	30. 6. 1961	Moteur électrodynamique	Electrodynamic drive
357 355	30. 9. 1961	Dispositif électromagnetique à aimant permanent permettant d'entretenir le mouvement d'un organe	Drive for a movement with a permanent magnet
360 030	31. 1. 1962	Dispositif de remontage et de mise à l'heure	Special form of winding and hand-setting mechanism
360 707	15. 3. 1962	Dispositif d'affichage	Electronic reduction drive
367 443	15. 2. 1963	Oscillateur mécanique	Mechanical oscillator
367 446	15. 2. 1963	Montre à remontoir automatique par masse oscillante	Automatic winding wrist watch with direct centre seconds (for Raymond Polo)
371 053	31. 7. 1963	Mécanisme de mise à l'heure pour pièces d'horlogerie	Simplified hand-setting mechanism (for Raymond Polo and Marcel Loichat)
375 290	15. 2. 1964	Montre à remontoir automatique par masse oscillante	Addition to Pat. 367 446, extension of bearing arbor (for Raymond Polo)
375 422	29. 2. 1964	Moteur à courant continu	Direct current motor to drive a timepiece (for Robert Favre)

List of patents granted to L.A. & I. Ditesheim and Movado (continued)

Patent No.	Dated	Original title	English title and remarks
385 745	15. 12. 1964	Oscillateur à torsion pour montre électronique	Oscillating element for electronic watches
387 552	31. 1. 1965	Dispositif électronique pour la stabilisation de la vitesse d'un moteur pour pièce d'horlogerie	Electronic stabiliser for a motor to drive a timepiece
389 023	15. 3. 1965	Compteur d'impulsions électriques bidirectionnel	Bi-directional electrical impulse counter
389 509	15. 3. 1965	Montre à remontage automatique par masse oscillante	Automatic winding for a watch with ball bearing rotor (for R. Polo)
397 540	15. 8. 1965	Montre à quantième	Watch with date indication
399 330	15. 9. 1965	Pièce d'horlogerie à remontoir automatique par masse oscillante	Automatic winding watch movement with the winding rotor train independently mounted (for R. Polo)
400 913	15. 10. 1965	Raquetterie pour mouvement d'horlogerie	Regulator index system
408 796	28. 2. 1966	Relais électronique sans contact mobile	Electronic relay without contacts
410 796	31. 3. 1966	Pièce d'horlogerie d'électronique	Electronic watch movement
414 472	31. 5. 1966	Pièce d'horlogerie comprenant un oscillateur mécanique	Movement with a mechanical oscillator
434 123	15. 4. 1967	Montre-bracelet électronique	Electronic wrist watch
447 049	15. 11. 1967	Pièce d'horlogerie électronique	Electronic timepiece
447 050	15. 11. 1967	Montre-bracelet électronique	Electronic wrist watch
452 443	29. 2. 1968	Oscillateur pour pièces d'horlogerie	Oscillator for timepieces
453 219	15. 3. 1968	Montre-bracelet de forme	Fixing of a form watch movement in the case (for Raymond Polo)
457 285	31. 5. 1968	Oscillateur à torsion pour montre électronique	Torsion oscillator for electronic watches
462 044	31. 8. 1968	Oscillateur à torsion pour pièce d'horlogerie	Torsion oscillator for timepieces
472 062	30. 4. 1969	Oscillateur mécanique pour montre électromécanique	Mechanical oscillator for electromechanical watches
481 411	15. 11. 1969	Resonateur de rotation mécanique pour appareil de mesure du temps	Mechanical resonator for electromechanical timepieces
483 045	15. 12. 1969	Dispositif de stabilisation de la fréquence d'oscillation d'un oscillateur mécanique pour appareil destiné à la mesure du temps	Device to stabilise the oscillation frequency of a mechanical oscillator for timepieces

Swiss case makers

The following tables show the punch marks of Genevan and other Swiss case makers that are to be seen on Movado watches.

These punch marks mostly consist of a particular form of indentation (key, hammer, etc.) together with a number, whereby the number – which must be registered at the Swiss Federal Institute of Intellectual Property in Berne – stands for the name of the case maker. The tables show the key to these numbers referring to the names of case makers.

Examples:

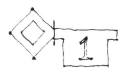 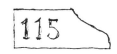

It can also happen that a case is signed with two initials. The Genevan firm François Borgel, for example, which made water-resistant cases for Movado in the 1940's and 1950's, used the following design:

Genevan case makers:

Number	Name
1	Wenger
2	Baumgartner
5	Croittier
20	Artisanor
26	Ponti, Gennari & Co.
28	Atelier réunit

Other Swiss case makers:

Number	Name
102	Amez-Droz
104	Grimm, Raoul
105	Blum, fils de Jules
107	Cattin, F.
108	Challandes, M.
109	Chatelquin-Sandoz
115	Favre et Perret
116	Flückiger & Co., St. Imier
117	Gunther & Co.
119	Gindraux & Co.
121	Guillod & Co.
122	Guyot, Alcide
123	Pfeniger
124	Ducommun, Fils de G.
126	Humbert & Co.
127	Schlaeppi & Co., Edmor
128	Jung & Fils
129	Junod & Co.
132	Monnier & Co.
136	Soillman & Co.
140	Dubois, B. C., Le Locle
141	Gabus, Frères, Le Locle
143	Nardin, Ch., Le Locle
147	Kohli, Chs., Tramelan
148	Vuillomet & Co., Bienne
149	Wyuss & Co., Granges
150	Bohlen, E., Granges
151	Hegendorn, H., Granges
152	Leuenberg & Co., Granges
158	Arnoux & Co., Le Noirmont
161	Erard, J.
163	Miserez, A. C., Saignelegier
166	Jeanneret, H., Le Locle
167	Carnal & Co.
170	Dubois, Chs. & Co., Le Locle
172	Monnier & Vaucher, Novelor
201	Bodemer & Aab
202	Bonnet & Cie.
205	Méroz, G., Neuchâtel
206	Bräuchi, A.
207	Niestlé, Peseux
210	Rubattel, Wayermann & Co.
213	Marcanti & Desfourneaux
214	Montandon, Frères, Le Locle
217	Lampert, J.
218	Gianoli, E.
219	Soguel, Frères
221	Paolini
270	Chs. Reinbold S. A.
350	S. S. B. O.
351	Graber, Percival
353	Stila S.

268

Movado wrist chronograph in a water-resistant steel case according to the FB Patent (François Borgel), made in 1955. Silvered dial with small seconds, 60-minute and 12-hour registers, signed "Movado", inscribed on the case back "Sub Sea", Reference No. 2387 95704 568. Nickel-plated 12''', Calibre 95 M movement with lever escapement, 17 jewels. The case back shows the impressions of the case maker's stamp as well as of the Movado manufacturer's stamp and the Case Reference No. 95 704 568 (used between 1966 and 1971).

269
8-day window display clock in a gilded hemispherical case, made in 1960. Dial and bezel signed "Movado Paul E. Morrison", the centre seconds hand beats full seconds. Quality finished three-quarter plate movement with lever escapement, for export, 24 jewels, swan-neck spring precision regulator index, 3 adjustments.

Museums in which Movado watches are to be found

Denmark: Museum of Decorative Art, Copenhagen
Germany: Kunstgewerbemuseum, Schloß Köpenick, Berlin
Deutsches Museum, Munich
Die Neue Sammlung, Staatliches Museum für angewandte Kunst, Munich
Museum Ludwig, Cologne
Finland: Finnish Watch and Clock Museum, Helsinki
Great Britain: Design Museum, London
Victoria and Albert Museum, London
Japan: Kawasaki City Museum, Kawasaki
Sezon Museum of Art, Tokyo
Columbia: Museo de Arte Moderno, Bogota
Netherlands: Museum Boymans-van Beuningen, Rotterdam
Austria: Museum für moderne Kunst, Vienna
Switzerland: Museé International d'Horlogerie, La Chaux-de-Fonds
Museum Altes Zeughaus, Solothurn
Uhrenmuseum Beyer, Zurich
Spain: Museo de Bellas Artes, Bilbao
USA: American Clock and Watch Museum, Bristol, Connecticut
Movado Company Store, New York
Venezuela: Museo de Arte Contemporaneo, Caracas

270
A valuable pocket watch with an enamelled crest, made about 1915.

Sources of illustrations

(The figures refer to the numbers of the illustrations)

Antiquorum, Geneva   152, 162

Archiv Pfeiffer-Belli, Munich   130

Archiv Süddeutsche Zeitung, Munich   259 c

Art & Work, Pforzheim   119–121, 266

Baur, Alexander, Munich   69

Baggenstoß, Toni, Solothurn   202

Cooper-Hewitt-Museum, New York   218, 220, 265

Dr. Crott Auktionen, Inh. Stefan Muser, Mannheim   128, 264

Deutsches Patentamt, Munich   8, 22–24, 33, 34, 75–77, 92, 106, 131, 146, 180, 181, 184, 187, 194

Ditesheim, Bernhard   6

Ditesheim, Gérard   2, 219

Dorotheum, Vienna   82

Editions d'en Haut (Hrsg.) La Chaux-de-Fonds XX⁰ siècle, La Chaux-de-Fonds 1993   1 a-b

Eells, Georges, Hedda and Louella   262 c

Institut Der Mensch und die Zeit (Hrsg.), Der Mensch und die Zeit in der Schweiz   1291–1991, La Chaux-de-Fonds 1991   5

Joray, Claude, Bienne   19, 20, 29, 64, 124, 127 b-c, 185 a, 259 a-b, 273

Leiningen, Boris   257 d

Musée international d'Horlogerie, La Chaux-de-Fonds   10, 40, 41, 43–46, 49, 54, 55, 90, 111, 116, 140, 170, 171

Observatoire Cantonal, Neuchâtel   118, 119, 216

v. Osterhausen, Fritz, Lüneburg   4, 107

Photo 2000, La Chaux-de-Fonds   213, 267

Privatbesitz   20, 29, 69, 83, 93, 118, 121, 125, 202

Preller, Heiko, Düsseldorf   169

Simonin, Antoine, Neuchâtel   126, 127 a

Sotheby's, London   258

Stoeber, Bernhard, Ramsey   84, 104

Studio George Mauro, Little Falls N.J.   12, 13, 14, 18, 21, 25, 26–28, 30–32, 35, 36–38, 39, 42, 47, 50, 51, 53–63, 65–68, 70–74, 78–81, 85, 89, 91, 93, 95–97, 101–103, 117, 122, 123, 132–134, 137–145, 147–151, 153–161, 163 b–168, 172–177, 179, 182, 183, 185 b, 186 a, 188–191, 193, 196–201, 203–208, 210, 212, 214, 222–227, 229, 230, 234–236, 238, 257 a-c, 260–263, 265.

Swissair Photo-Vermessungen, Regensdorf   3

Vuille, Louis F., La Chaux-de-Fonds   213

Bibliography

Balfour, Michael, The Classic Watch

Barrelet, Jean Marc, Petit guide pour servir l'histoire de l'horlogerie, Neuchâtel 1988

Berman, Phyllis, Is traveling well the best revenge? in: Forbes August 1988

Bowman, Bernard U. jr., "Hermetic" Watches: a review, in: Bulletin of the National Association of Watch & Clock Collectors (NAWCC Bulletin), Vol. 36/3 No. 290 June 1994

Brunner, Gisbert L. and Christian Pfeiffer-Belli, Schweizer Armbanduhren, Munich 1990

Brunner, Gisbert L., Armbanduhren, Munich 1990

Brunner, Gisbert L., Basel von A bis Z, in: Alte Uhren 3/1988

Brunner, Gisbert L. and Christian Pfeiffer-Belli, Armbanduhren (Battenberg Antiquitäten-Kataloge), Augsburg 1992

Cleves, Charles, Movado – Ahead of its time, in: Horological Times November 1988

Cleves, Charles, Unusual watches from the 20s and 30s, in: Horological Times September 1989

De Carle, Donald, Complicated watches and their repair, in: Horological Journal July 1953 und May 1954

Dunning, Deanne Torbert, Interviews with Gerard Ditesheim, Gedalio Grinberg and Efraim Grinberg, unpub. ms. 1995

Eder, Norbert, Beobachtungsuhren, Munich 1987

Ehrhardt, Roy, Pocket Watch Price Guide, Kansas City 1972

Ehrhardt, Sherry und P. Planes, Vintage American and European Wrist Watch Price Guide, Kansas City 1984

Fabricants suisses d'Horlogerie (Ed.), Offizieller Katalog der Ersatzteile der Schweizer Uhr, 2 Bände, Soleure/La Chaux-de-Fonds 1949

Favre, Maurice, Daniel Jean Richard 1665-1741, Edition du Château des Monts, Le Locle, undated

Flume, Rudolf, Der Flume-Kleinuhr-Schlüssel K1, West Berlin and Essen 1958

Flume, Rudolf, Der Flume-Kleinuhr-Schlüssel K2, West Berlin and Essen 1963

Flume, Rudolf, Der Flume-Kleinuhr-Schlüssel K3, Berlin 1972

Fried, Henry B., Self-winding watches, in: Horological Times February 1990 and April, May, June 1991

Friedman, Richard, Watch-ing profits at $1 million a pop, in: New York Post of 9 September 1986

Golay, Hector, La Vallé de Joux de 1860 à 1890, Charbonnières 1979

Hampel, Heinz, Automatic Armbanduhren, Munich 1992

Horology Vol. VII April 1940 (author unknown), Waterproof watches

Humbert, B., Modern Calendar Watches, in: Journal suisse d'Horlogerie et de Bijouterie 1956

Humbert, B., Der Chronograph – Funktion und Reparatur, La Conversion 1990

Institut der Mensch und die Zeit (Hrsg.), Der Mensch und die Zeit in der Schweiz 1291–1991, La Chaux-de-Fonds 1991

Jaquet, Eugène and Alfred Chapuis, Histoire et technique de la montre suisse, Basel/Olten 1945. American edition (Technique and History of the Swiss Watch), New York 1970

Journal suisse d'Horlogerie et... (Ed.), Le Livre d'Or de l'Horlogerie, Geneva and Neuchâtel 1926

Kahlert, Helmut, Richard Mühe and Gisbert L. Brunner, Armbanduhren, Munich 1990

Kapteina, Thomas, Movado: It's "Bill-time", in: Armbanduhren International, No. 3 June 1993

Lang, Rüdiger und Reinhard Meis, Chronographen Armbanduhren, Munich 1992

Mann, Helmut, Die Cadraturiers im Joux-Tal, in: Alte Uhren 1/1984

Mann, Helmut, Gangleistung und Schwingungszahl, über die Bedeutung der Gangregler mit erhöhten Frequenzen, in: Alte Uhren 2/1986

Movado (Hrsg.), Uhren-Modellkatalog 1910–1920 (Fotosammlung)

Movado (Hrsg.), Uhren-Modellkatalog 1921 (Fotosammlung)

Movado (Hrsg.), Fabriques Movado, Switzerland (1948)

Movado (Hrsg.), Allo! Movado (Hauszeitschrift), Nrn. 7/1962 und 12/1967

Movado (Ed.), (Dunning, Deanne Torbert and Chuck Davidson), Andy Warhol Times/5 by Movado, 1988

Movado (Ed.), (Dunning, Deanne Torbert and Chuck Davidson), The Legend behind the Museum Watch, undated

Movado (Hrsg.), Diverse Broschüren über die einzelnen Künstleruhren, die Bauhaus-Uhr, die Soldatenuhr und die Collection 1881, Grenchen 1988–1994

Neumüllers, Herbert, Die Observatoriums-Serien von Movado 1912-1939, Unpub. ms.

Osterhausen, Fritz von, Armbanduhren Chronometer, Munich 1990

Osterhausen, Fritz von, Movado – immer in Bewegung, in: Alte Uhren 3 und 4/1991. Translation into English by Bernhard Stoeber, in: Horological Times April, May und June 1993

Perrenoud, Marc, Un rabbin dans la cité – Jules Wolff, Neuchâtel 1989

Pritchard, Kathleen H., Fabriques Movado, in: Encyclopedia of Swiss timepiece manufacturers, unpub. ms.

Rohr, Frank, Gedalio Grinberg – discreet Tycoon of North American Watch Company

Sexton, Norma, Making Movado, in: Modern Jeweler February 1988

Sotheby's London (Ed.), Auction catalogue of 15.12.1992

Thomann, Charles, Les dignitaires de l'horlogerie, Neuchâtel 1981

Thompson, Joe, Movado celebrates its centennial 1881–1991, in: Jewelers' Circular Keystone February 1981

Viola, Gerald und Gisbert L. Brunner, Zeit in Gold, Armbanduhren, Munich 1988

Zigliotto, Eugenio, Movado: A make always on the move, in: International Wristwatch, USA Edition, No.17/1992

Index

The numerals in normal type refer to the text pages, the
figures in italics to the numbers of the illustration.

Agam, Yaacov   168, 170
Arman   172
Art déco watches   67, *80, 87, 89–91*
Artists' Watches   168, 170, 172, 174, 176, 178
Aubert, Daniel   8
Audemars, Louis   8
Audemars Piguet, Fa.   15, 18, 46
Augsburger, Louis   98
Automatic watches   128, 133 f., 136, *178–212*

Bagnolet calibre   38
Bayard, Thomas   7
Bell, Alexander Graham   9
Berlin, Irving   190
Biaudet, Mathieu   8
Bill, Max   176
Bismarck   9
Blind users, watch for   *97*
Bonaparte, Napoleon   26
Borel, Ernest   170
Borgel, François, Co.   104, *147, 172, 268*
Boris III, King of Bulgaria   188, *257*
Brandt, Louis   18
Braunschweig, Alphonse   10, 18
Breguet, Abraham-Louis   8, 91
Breguet balance spring   *128, 130*
Breitling, Co.   156
Breitling, Léon   18
Breuer, Marcel   148
Britto, Romero   178
Bulova, Co.   18
Burki, Kurt   167

Calder, Alexander   170
Calendar wrist watches   124, 126, *162, 170–177*
Calvin   7
Cartier, Co.   86
Case makers   228
Castella, Paul   160, 165
Castro, Fidel   164 f.
Catalogues   39, 46f., *40, 41, 43–46, 49*
Champlevé technique   *86, 87*
Chapuis, Alfred   50, 86
Chronograph
– Single push-button C. 46, 120, *40, 64, 167*
– Double push-button C. 104, 118, *145*
Chronograph with split seconds   46, 118, *48*
Chronometer   46, 90 f., 93, 95 f., 98
– Calibres and series 93, 95 f., *196–214*
Chronometer trials   48, 90 f., 98, 144, 146 f., *216–214*

Churchill, Sir Winston   188, *258*
Citizen, Co.   165
Cleves, Charles   38
"Clinergic 21"-escapement   6
Cloisonné technique   86, *96*
Collections, present-day   184
Concord, Co.   165, 167
Conway, Alban   190, *261*
Corum, Co.   165
Cottier, Louis   134

Daniels, Greta   148, 150
De Carle, Donald   128
Didisheim, Marc and Emmanuel   10, 18
Digital watches   *148–151*
Ditesheim, Co., see also Ditesheim, Achille   14 ff., *7, 8*
Ditesheim, L.A. & I., Co.   16, 18, 20, 90, *8, 9*
Ditesheim, L.A.I. & Frère, Co.   23, 26
Ditesheim, Family   7, 9 f., 12
    Ditesheim, Abraham   10, 12, 23
    Ditesheim, Achille   10, 12, 14 f., 16, 26, 89, 128, 154,
    180, *4, 6*
    Ditesheim, Alain   154
    Ditesheim, Armand   46, 154
    Ditesheim, Aron   10, 22
    Ditesheim, Bernard   6, 154
    Ditesheim, Bertin   154
    Ditesheim, Edmond   98
    Ditesheim, Edouard   154
    Ditesheim, Gaston   89
    Ditesheim, Georges   89, 154
    Ditesheim, Gerard   128, 165
    Ditesheim, Isaac   10, 12, 22 f., 39, 89, 154, 165
    Ditesheim, Isidore   10, 12, 16, 34, 48, 89, *2*
    Ditesheim, Léopold   10, 12, 16, 18, 22
    Ditesheim, Louis   10
    Ditesheim, Lucien   46, 154
    Ditesheim, Pierre   89, 154
    Ditesheim, Roger   46, 128, 154
    Ditesheim, Samuel   9 f., 12, 16
    Ditesheim, Thérèse   10
Ditisheim, Family   10
Ditisheim, Paul   10, 12, 18
Dixi SA, Co.   160, 165
Doxa, Co.   18
Dreyfuss, Moise   10
Dubois, Werner Albert   98, 146
Dubois Depraz, Co.   156
Ducommun, Georges   18
Dutran, E.   18

Election, Co.   10, 18
Enicar, Co.   52
Engine turning   86, *96*
Enila, Co.   10
Eterna, Co.   18, 47, 50
Excelsior Park, Co.   18

Favre-Bulle, Frédéric-Louis   8
Favre-Jacot, Georges   18
Fernandez, Armand Pierre, s. Arman
Francillon, Ernest   18
Frey, Ernest   98
"Futuramic"-Calibre   133, *181, 184, 187, 196*

Garland, Judy   190
Girard-Perregaux, Constantin   18
Girard-Perregaux, Co.   50
Golay, Georges   8
Grandjean, Henri   8
Grinberg, Efraim   164, 184, *240*
Grinberg, Gedalio   164 ff., 168, 178, 184, *240*
Gropius, Walter   148
Gruen, Co.   38
Guillaume balance   93, *117, 119, 123, 125, 126, 128*
Guye, Henri   147

Hänggeli, Charles   98
Hamilton-Büren, Co.   156
Hansen, Gerry   160
Hart, Moss   190, *261*
Haskell, Douglas   150
Hellinger, Edwin   89
Hemingway, Ernest   165
Hermética SA., Co.   76, 86
Heuer, Edouard   18
Heuer, Co.   18
Heuer-Léonidas, Co.   156
Hirsch, Achille   10
Hitler, Adolf   188
Hopper, Hedda   190, *262*
Horwitt, Nathan George   148, 150 f., 160, 165 f., 182,
    218, 219, 220, 253
Houriet, Jacques-Frédéric   8
Huguenin Frères, Co.   74

Ingersoll, Co.   52
Itten, Johannes   170
IWC, Co.   46, 50

Jaeger LeCoultre, Co.   34
Jaquet, Eugène   50, 86
Jeanneret, J.F.   18
JeanRichard, Daniel   7 f., 11
Jenkins, Paul   168
Jürgensen, Family   8
Jugendstil, see Art déco watches
Junghans, Co.   176

Kaufmann, George S.   190
Krähenbühl, Rémy   154

Lam, Wilfredo   168
Laurent   18
LeCoultre, Charles-Antoine   8
LeCoultre, Co.   14, 150
Le Phare, Co.   160, *12, 14*
Lépine Calibre   52
Leroy, Co.   20
Lévy, Salvator   16
Longines, Co.   18, 50, 147, 150
Luxor, Co.   160

Marvin, Co.   10, 18
Meylan, Philippe-Samuel   38
Meylan, Samuel-Olivier   8
Mies van der Rohe, Ludwig   148
Mistral, Frédéric   16
Models
    "Acvatic"   104, 128, *142, 149*
    "Andy Warhol Times/5"   168, *243*
    "Apogée"   18 f., 90, *11*
    "Astronic"   134, 203, 227
    "Astronic HS 360"   156, 160
    "Bill-time"   176, *249*
    "Black Sapphire"-Collection   184, *255*
    "Calendermeto"   86, 124, 126, *97, 103*
    "Calendette"   *173*
    "Calendograph"   108, 118, 124, *172, 258*
    "Calendolette"   124
    "Calendolux"   126, 133
    "Calendomatic"   124, 126, 133, *174*
    "Calendoplan"   126, 133
    "Calendoplan Automatic"   126, *175*
    "Calendoplan Baby"   124
    "Calendoscope"   *195*
    "Celestograph"   124, *162, 171, 176, 177*
    "Centenaire"   160, *237*
    "Chronodiver"   158
    "Chronograph"   108, 118, 121
    "Collection 1881"   167, 180, *251*
    "Cronacvatic"   104, 120
    "Cronoplan"   104, 108, *145, 146*
    "Curvex" (Gruen Co.)   38
    "Curviplan"   104, *131, 133–136, 140, 141*

"Cyclox"   148, *218*
"Datachron"   158
"Datron"   158, *230*
"Datron HS 360"   156, *228, 229*
"Delirium"   160, 165
"Duoplan" (Jaeger LeCoultre Co.)   34
"Elapse, Eclipse, Ellipse"   174, *248*
"El Primero"   120, 156, 158, *228*
"Ermeto"   6, 39, 74, 76, 86, *92–115*
"Ermeto Bag"   86, 104, *101*
"Ermeto Calendine"   86, 124, 93, *114*
"Ermeto Luxe"   86, *104*
"Ermetophon"   76, 86, *93, 113, 115*
"Ermeto Pullman"   76, 86, *93, 112*
"Galaxy"-Collection   170, *245*
"Horizon"-Collection   154, *256*
"Kingmatic"   133 f., 136, 144, 158, *190, 197, 217*
"Kingmatic Calendar"   133
"Kingmatic Calendoplan"   134
"Kingmatic Chronomètre"   134, 136, *215*
"Kingmatic HS 360"   136, 144, 154, 158
"Kingmatic S"   136, *214*
"Kingmatic Sub Sea"   136, *197*
"Livre d'Heures"   *153*
"Love Star"-Collection   170, *244*
"Movascope"   126, *175*
"Museum Imperiale"   160, *239*
"Novoplan"   104, *137, 139*
"Polygraph"   134, *193, 194*
"Polyplan"   6, 27, 34 f., 38, 104, *33, 34, 35, 37, 38, 39*
"Queenmatic"   133, 158, *188*
"Rainbow"-Collection   170, *246*
"Ralco"   46, 118, *48*
"Reverso" (Jaeger LeCoultre Co.)   34
"Reverso" Movado type   108, *152*
"Sunwatch"-Collection   184, *256*
"Tempomatic"   128, 133, 158, *178, 179*
"The Children of the World"   178, 250
"The color of time"   172, *247*
"Valentino"   66, *75–79, 81, 82*
"Video"   158, *233*
Moholy-Nagy, Laszlo   148
Movado watch owners   188, 190
Movado-Zenith-Mondia (MZM)   158
·Movement calibres and movement numbers   47 f., 195–221
Museum stocks   230
Museum Watch   148, 150 f., 160, 166, 182, *219–221, 231, 232, 236, 238, 239, 253, 254*

Nardin, Ulysse   8, 18
Neumüllers, Herbert   98
Nevelson, Ruth   168
Niello technique   *85, 97*
North American Watch Corporation (NAWC)   165 ff.

Omega, Co.   18, 46 f., 50

Pannett, Michael J.   160
Parsons, Louella   190, *262*
Patek Philippe, Co.   134
Patents   22–27, 38, 222–227, *22, 23, 24, 33, 34, 75, 76, 77, 106, 107, 146, 180, 181, 184, 222–227*
Paul Buhré, Co.   160
Pendant watches   66, *79, 81, 82*
Peron, Evita   128
Peron, Juan   128
Perrenoud, Marc   10
Pery Watch Co.   10
Peseux, Co.   147
Phillips terminal curve   *93, 117, 119, 123, 125, 126*
Piaget, Co.   165
Piccard, Auguste   188, 190, *259*
Piguet, Frédéric   108, 118
Piguet, Louis Elisée   8, 108
Piguet, Victorin   8
Polo, Raymond   154
Porter, Cole   190

Railway employees' pocket watches   16, 18
Ramolayé technique   23
Reproductions   182
– Bauhaus Watch   182, *253*
– First World War Soldier's Watch   182, *252*
Rolex, Co.   47
Rosenquist, James   174
Roskopf, Georg Friedrich   8
Rotary, Co.   10

Sampras, Pete   166, *241*
Sand, George   9
Sandoz, Abraham-Louis   11
Sautebin, Georges   147
Schild, Urs   18
Schmid, Charles-Léon   10
Seiko, Co.   165
Signatures
    "Alarm 8 Days"   *20*
    "Brock and Co."   *142*
    "Bruxelles 58"   *103*
    "Cartier"   *68, 97, 261*
    "Chiswell Hnos"   *13*
    "Chronometer Movado"   *35, 38, 70, 133, 134*
    "Chronomètre à Ancre"   *29*
    "Chronomètre de Bord"   *128*
    "Chronomètre Movado"   *27, 51, 56, 62, 63, 65, 67, 74, 78, 81, 82, 91, 96, 97, 127, 132, 144, 186, 213, 214, 217, 257, 259*
    "Chronomètrè Movado Suiza"   *54*
    "Chronometro Escasany"   *46*
    "Coincidence"   *116*

"Contador" *25*
"E. Gübelin Lucerne" *138*
"F. Lastretti Bahia-Blanca" *32*
"Gübelin" *148*
"Mistral" *46*
"Movado" *31, 42, 53, 54, 56, 58, 59, 60, 61, 64, 66, 68,*
  71, 72, 73, 83, 84, 86, 88–90, 116, 132, 134, 137,
  139, 140, 143, 145, 146, 153, 157, 158, 159, 160,
  161, 163, 164, 165, 167, 172, 176, 177, 189, 193,
  196, 199, 204, 205, 206, 207, 208, 210, 212, 223,
  224, 226, 258, 262, 268
"Movado 8 Days" *21*
"Movado Automatic" *185, 191*
"Movado Beyer" *173*
"Movado Cartier" *102*
"Movado Cronoplan" *145*
"Movado Factories" *36, 112–114, 123, 125, 126, 133,*
  147, 164, 167, 172, 173, 177, 182, 183, 185, 186,
  188, 190, 193, 197, 206, 207, 212, 213, 214, 227,
  229, 234, 235, 257
"Movado Fy" *21, 28, 53, 55, 65, 70, 71, 117*
"Movado Non magnetic" *69, 166, 169*
"Movado Paul E. Morrison" *269*
"Movado Solidograf" *201*
"Movado Sport Non-magnetic" *200*
"Movado Suiza Carlos u. Perret Rosario" *95*

"Movado Sûreté" *12, 26, 28, 31, 85*
"Movado Watch Fy, Sûreté" *55, 119*
"Movado Zenith" *238*
"Recuerdo de Empleados De Sun Life en Guatemala
  Agusto 1930" *73*
"Rosenberg-Wallach, Joyeros, Lima Peru" *46*
"Sûreté Suiza" *25*
"Tanit" *46*
"Tiffany & Co." *101, 158, 168, 188, 198, 225*
"UTI Paris-Movado-Spitzer & Fuhrmann" *222*
Soldier's Watch, First World War 50 ff.
Solvil, Co. 18
Special models *153–158*
Steichen, Edward 148
Streitberger, Gerhard 27

Tauber, Brüder 104
Tiffany & Co., Co. 86
Timex, Co. 165
Tinguely, Jean 170
Tissot, Charles-Frédéric 18
Tourbillon 93
Trey, César de 76

Ulysse Nardin, Co. 52, 98, 147
Universal, Co. 156

Vacheron & Constantin, Co. 148, 150
Valentino, Rudolph 66, 89
Van Cleef & Arpels, Co. 165
Vasarély, Victor 170
Vermeil technique *97*
Victoria, Queen 9

Wagner, Henri 154
Waltham, Co. 52
Warhol, Andy 168, 190, *260*
Watson, Harry 190, *262*
Weiss, Fabian 165
Weiss, José 165
World Exhibitions, prizes won at 20
Wolff, Jules 128
Wrist watch chrongraphs 118, 120, *160, 162–169*
Wrist watch chronometers 144, 146f., *213, 214, 216,*
  217

Zamenhof, Ludwig 23
Zenith, Co. 18, 47, 86, 120
Zenith-Movado, Co. 144, 156, 158, 160
Zenith Movado Le Locle SA, Co. 160, 165
Zenith Time SA, Co. 160
Zodiac, Co. 160

L.A.I. DITESHEIM   L.A.I. DITESHEIM

MISTRAL   MOVADO   POLYPLAN

RALCO   VALENTINO   ERMET

ERMETOPHON   CURVIPLAN   NO

CRONOACVATIC   CHRONOGRAF

CALENDOPLAN   MOVASCOPE

CALENDERMETO   CALENDACV

QUEENMATIC   KINGMATIC

CALENDETTE   CALENDINE   MUS